Behavioral Enrichment in the Zoo

Behavioral Enrichment in the Zoo

HAL MARKOWITZ
Department of Biological Sciences
San Francisco State University

VAN NOSTRAND REINHOLD COMPANY
NEW YORK CINCINNATI TORONTO LONDON MELBOURNE

Van Nostrand Reinhold Company Regional Offices:
New York   Cincinnati

Van Nostrand Reinhold Company International Offices:
London   Toronto   Melbourne

Library of Congress Catalog Card Number: 81-10309
ISBN: 0-442-25125-4

Manufactured in the United States of America

Published by Van Nostrand Reinhold Company Inc.
135 West 50th Street, New York, N.Y. 10020

Published simultaneously in Canada by Van Nostrand Reinhold Ltd.

15 14 13 12 11 10 9 8 7 6 5 4 3 2 1

Library of Congress Cataloging in Publication Data

Markowitz, Hal.
   Behavioral enrichment in the zoo.

   Includes index.
   1. Zoo animals—Behavior. 2. Zoological gardens—
Design and construction. I. Title.
QL77.5.M38   712'.5   81-10309
ISBN 0-442-25125-4   AACR2

To Krista, Tim, and Jenny

PREFACE

It is impossible to begin to properly thank or even mention by name the hundreds of colleagues, students, volunteers, and zoo staff members who have helped in the work described in this book. The reader will more frequently find me using the term *we* than *I* as I describe various projects, because they almost all resulted from mutual planning and hard work by a number of people. In particular, I do want to thank all of the following without whose help little that is described here would have ever been accomplished: Mike Andrews, Barbie Barrow, Helene Burgess, Russ Chevrette, Ian Curington, Ron Fial, Franci Files, Mary Garrison, Jay Haight, Roger Henneous, Brian Johnson, Nick Lee, Phil Lipton, Ann Littlewood, Rob Mattson, Mike McBride, Jill Mellen, Marilyn Mink, Anne Moody, Tom Moore, Marilyn Morrow, Leonie Nadal, Phil Ogilvie, Diane Roelandt, Mike Schmidt, Georgianne Schmuckal, Les Squier, Sue Stanley, Vic Stevens, Olan Vineyard, and Wilfried Zimmerman.

I hope it won't be too disconcerting for the reader that occasionally my personal opinions and judgments are intertwined with more data-based outcomes. Where this occurs, I have tried to clarify which ones are empirical statements and which ones are theoretical assertions. Inevitably, in spite of my trying repeatedly to indicate that the mechanistic aspects of our procedures were designed to increase rather than decrease the domain of possible behaviors in zoo residents, there will be some who find the use of mechanical and electronic technology unpalatable. For those who are not dissuaded from that position by things they may read in this book, I would offer a friendly, rather than avaricious, challenge: find methods to their own liking to enrich the desolate captive environments in which most of our zoo animals live.

It is easy to see why some critics have become so polarized and often nonconstructive in their evaluations of zoos. These institutions control the lives of wonderful animals and the very existence of some species. Zoo animals deserve much more intensive efforts to improve their captive environments and opportunities for meaningful activities. The premise throughout this book is that zoos, aquariums, and other wildlife facilities are irreplaceable resources badly in need of constructive criticism and help in progressing as more humane and naturalistic institutions. This progress will not be furthered by categorical condemnation or praise.

On balance, I see hopeful signs in an increasing percentage of zoo staff, administrators, and governing bodies that are sensitive to the need for more than lip service and window-dressing improvements. All of us as citizens, with collective responsibility for institutions in which we have placed other living creatures, need to work constructively to accelerate the rate of improvement in zoological procedures. A final personal prejudice that I have tried to expand upon in several sections of the book is that in new planning, the resident animals must come first. Rather than excessively compelling zoo residents to learn new behaviors for our amusement or edification, the work described herein was initiated to provide a chance to reduce the monotonous, sterile routine to which the majority of animals in captivity are ordinarily limited.

Occasionally, people have too kindly attributed the idea that zoos can profitably use technology to improve the plight of captive animals to our research group. The truth is, of course, that long before any of us were born, there were people arguing for the need for such improvements. In retrospect, I regret that this book continues the tradition of largely unacknowledged debts to many such people. Perhaps that should form the basis of another work, but, without meaning to slight others, I must pay special tribute to the influence of the writing of Heini Hediger, an inspiration to any person who has ever considered working with captive animals.

HAL MARKOWITZ

CONTENTS

Behavioral Enrichment
in the Zoo

1
INTRODUCTION

Increasing behavioral opportunities for captive animals is hardly a new idea. The need for this has certainly been felt by many laboratory workers who find the barren cages in which their subjects live the least palatable part of experimental work. The effects of rich versus barren environments have, in fact, been the focus of a great number of studies (Denenberg, 1962; Meyers, 1971; Mitchell, 1973). These studies have related environmental enrichment to parameters ranging from enhanced behavioral capability through changes in brain structure (Diamond, Krech, and Rosenzweig, 1973; Meyers, 1971; Tapp and Markowitz, 1963).

Robert Yerkes, whose primate work has served as a model for many contemporary investigators, states the case clearly (Yerkes, 1925):

Undoubtedly, kindness to captive primates demands ample provision for amusements and entertainment as well as for exercise. If the captive cannot be given the opportunity to work for its living, it should at least have abundant chance to exercise its reactive ingenuity and love of playing with things.

The greatest possibility of improvement in our provision for captive primates lies in the invention and installation of apparatus which can be used for play or work.

The dilemma that faces every sentient individual becomes increasingly clear when the experience of a zoo visit is carefully examined. Opportunities to visit and take your children to see magnificent creatures, many of which one otherwise might never encounter in a

1

lifetime, is an exciting recreational experience and can be educationally powerful. However, in the majority of cases, the character of the captive environment leaves so much to be desired that it brings an uncomfortable feeling of similarity to traditional human prisons.

This last paragraph should not be interpreted as a condemnation of zoos or of their personnel. Inadequacy in traditional zookeeping is one focus of attention at almost every professional zoo meeting. The husbandry efforts of zoos have led to the preservation of some species that have become extinct in the wild (Hediger, 1950). Modern zoo workers have also been leaders in the preservation of wild habitats and in nature conservation and have encouraged governmental protection of endangered and threatened species. These are only a handful of the important contributions of well-run animal conservatories.

This book takes as a point of departure that there are many good reasons that we should maintain zoos and evolve them into the best possible facilities. *Best* means: 1. the best possible home for animals that have been removed from their natural habitat, 2. the best educational recreational experiences for zoo visitors, and 3. the best provision for research of all kinds beneficial to the resident species. That zoos do not yet begin to approach these ideals means that they need constructive help from all sorts of animal lovers, and especially from those willing to give time, energy, and thoughtful consideration to practical methods of improvement.

One of the revolutions in zoo exhibits has been the movement toward *naturalism* which will be a major topic of the next chapter. This revolution has included the development of some magnificent wildlife parks and the increased use of real and fiberglass foliage in zoological gardens. Unfortunately, this advance in zoo concepts has seldom been paralleled with a *careful* analysis of the behavioral contingencies in nature that may be missing in captivity (Markowitz, 1976; Markowitz and Stevens, 1978; Schmidt and Markowitz, 1977). From the standpoint of the zoo visitor, it increases excitement to see an animal in a naturalistic setting, and it certainly contributes to the educational experience when plants and terrain similar to those native to the animal's home in nature can be included. However, much of the beauty of most species involves their *behavior*, not just their "backdrop," coloration, and physical features. Seeing a serval pounce upon ground prey or flush gamefowl from the bush and snare it on the fly cannot be paralleled by placing these beautiful African cats in an inanimate exhibit, no matter how lush the exhibit. Changes in feeding procedures and other behavioral protocol must accompany the new artwork, landscaping, and architecture.

The zoo animal deserves our primary attention since its comfort and health are involved every day, while the average visitor attends the zoo a few times a year at most. From this perspective, the need for behavioral opportunities is even more apparent. What could be more unhealthy for most species than an unresponsive environment where their sustenance requirements are met on some arbitrary schedule by their keepers? The effects of such institutional schedules on humans is well known, and their effects on animals other than *Homo sapiens* must not be underestimated. The technology exists for an almost endless variety of methods to provide animals some control over their own schedules of feeding, drinking, and exercise. This book deals with some of the first concentrated efforts to apply this technology to the design of zoo exhibits.

Contrast the appearance of the serval in Figures 1-1a and 1-1b and of the bear begging for food versus diving to obtain some fish (Figures 1-2a and 1-2b). These show what can be accomplished with environmental enrichment. For the great majority of species, a survey of the field literature can provide a blueprint for the development of exciting homes for our animals. These in turn will attract increasing numbers of people to visit and learn about the animals' natural abilities.

Along with the excitement of developing new plans and testing prototypes come significant lessons in humility. Animals will often find ways to radically alter the designer's conceptions of the way apparatus is to be utilized. In Chapter 12, there are many examples which show that other species can outsmart humans. False starts resulting from the animal's inventive responses are obviously frustrating at the time they occur, but if one can keep a sense of humor, they are one way of letting the animals teach us a great deal.

Everyone who has lived or worked closely with animals knows how individualistic they are. One of the most rewarding aspects of developing increased behavioral opportunities is watching the expression of these individual differences. Given opportunities to eat in exchange for a little work whenever they choose, some monkeys will exploit others, some will find cooperative ways to complete food-gathering

Figure 1-1a. Typical day-long behavior of a serval *(Felis serval)* in the zoo. *(Photo by W. Wong)*

tasks, and some will be iconoclastic, insist on earning everything themselves, and never share. Although quantification is a great challenge when studying the behavior of animals in groups, the richness of these observations is unparalleled in traditional testing situations, where individual organisms are typically isolated. Watching the harbor seals arrange conventions for who will work and who will eat (Chapter 7), and observing young animals learn how to earn food by watching their elders brings some of the excitement of field research to the zoo.

Behavioral enrichment in the zoo is in its infancy and as it grows and becomes integrated with the broad scope of modern zoo design, it will bear little resemblance to the crude early beginnings described in much of this book. Lights, levers, and games will be replaced with

Figure 1-1b. Serval leaps to capture artificial prey. *(Photo by B. McCabe)*

naturalistic materials that contribute to the zoo's educational function. In a paper entitled, "In Defense of Unnatural Acts Between Consenting Animals" (Markowitz, 1975), I described some analogies between the human condition and that of other, institutionalized species. All of us have come to recognize the value of developing entertaining methods of exercise now that we no longer typically need to gather our own food or actively defend our families. Bicycles that do not

Figure. 1-2a. Polar bear *(Thalarctos maritimus)* begs for food. *(Photo by H. Markowitz)*

move and rowing machines that go nowhere are regularly prescribed to improve the human physical condition. Animals in zoos and other

Figure 1-2b. Polar bear dives for fish which bears have "ordered" (see Chapter 6).
(Photo by J. Mellen)

captive facilities, which are insulated from the requirements of the wild, also need activity programs to remain healthy. At first blush, it is appealing to think that we are doing wonderful things for a species by asking nothing of them except procreation and appreciation of a safe environment, but there are few of *us* who would remain happy for very long under circumstances where we had no control of our own life contingencies.

Given a choice, animals will typically work to obtain food even when the same food is available free (Neuringer, 1969; Stevens, 1978). This laboratory finding, which has now been replicated often enough to be accepted by most behaviorists, was initially astonishing to some. There remains a tendency for us to think of animals as unfeeling automata, but anyone who has worked closely with them knows that this is foolish. Our language, which has been so influenced by theological and Cartesian traditions, helps to reinforce and perpetuate an excessive dichotomy between humans and other animals (Beach, 1955). Within the behavioral sciences, the tautological use of the term "instinct" has also added to the problem (Lehrman, 1953). It is clear that many theoreticians are unwilling to accept the fact that humans are responsive to environmental contingencies and past histories of reinforcement in the same fashion as other animals. In the most ludicrous modern terminology, we see the same degree of plasticity in behavior referred to as "insightful learning" in humans and as "learned instincts" in other species.

The discovery that animals would work for food in the presence of identical, free food has also been disquieting for some traditional learning theorists. After all, it illustrates how glib and superficial some of our explanations of animal behavior are. A student coming to the laboratory and watching a rat press a lever for the first time or a pigeon peck a key is easily satisfied when you explain that they do this because they are hungry or thirsty and get food or water for their efforts. But, if they work to get food when standing next to a pile of food, this easy explanation becomes vapid. Instead, we are emphatically confronted with the proposition that other animals besides ourselves like to do things, to see things change because of their efforts, to enjoy the pride of gathering their own food or drink, and to have some control over their lives. This is what behavioral enrichment is all about.

2

ON NATURAL ZOOS AND UNICORNS

Throughout most of this book the term *naturalistic* rather than *natural* will be employed to describe efforts to reduce artificiality in the captive environment. This distinction is a critical one, since there are very few captive situations where managers would honestly want to return to a state of nature. At the same time, there is some tendency for promoters and public relations experts to maintain that zoos, wildlife parks, and aquariums are becoming natural. This is attractive to the public, partly because there seems to be a collective sense of guilt about having removed wonderful creatures from their homes in the wild. This chapter will suggest that it is possible, in some cases, to *improve* upon nature. *Nature* has become so confused in our usage that it is often implicitly synonymous with *God, pure,* and *good.* It is sad to reflect upon, but if nature requires *purity,* there is actually little nature left in the world.

The implausibility of natural zoos becomes apparent if we take a brief look at some of the best reasons for maintaining captive wild animals. First, there are programs specifically set up to reintroduce rare and endangered animals to the wild. In the wild, these species are often endangered because of the encroachments of civilization and general exploitation by humans. As their numbers dwindle, they may become increasingly susceptible to attack and destruction by their natural enemies. For some animals, zoos become sanctuaries and provide relative safety from all predation, both natural and man-made. There is a delicate balance that must be met in fairness to rare animals. On the one hand, it would be foolish to expose them to great danger. But, at the same time, it is unfair to remove the stimulation of the wild environment without actively seeking new ways to provide a responsive milieu.

The Siberian tiger *(Panthera tigris altaica)*, the largest of the cats, is increasingly endangered and faces extinction in the wild. In zoos, this species has been so prolific that it is currently difficult to find a placement for new offspring. In the Portland Zoo, we watched a beautiful young pair of Siberian tigers grow to maturity and reproduce offspring. Their captive environment included some logs and vegetation and a pool in which they romped on occasion. But real excitement came at those infrequent times when a sparrow would chance too close to the grotto and be captured by these swift cats, or snow would come for the tigers to play in. Here is a species ripe for increased behavioral opportunity. This would in turn provide zoos a dynamic means to educate visitors about the capabilities of these animals.

Unlike most cats, the tiger is an excellent swimmer. Seeing a Siberian tiger, which may weigh as much as 290 kilograms (640 pounds) and be 4 meters (13 feet) long overall, race through the water is pure excitement. Since tigers tend to stalk and spring on their prey rather than undertaking long chases, it does not take great acreage to provide for naturalistic capture opportunities. In Chapter 10, some behavioral engineering for the Sumatran tiger is described. I hope that someday soon a progressive zoo director will plan a program for the Siberian tiger, a species so abundant in zoos that a decent prototype might encourage the development of active environments for many animals.

A second major possibility for "improving upon nature" in the zoo involves parasite loads and other veterinary considerations. It is not uncommon for disease and infestation to eliminate many animals in nature. There was a time when zoos, with mixed menagerie concepts and fragmentary beginnings of veterinary procedures, experienced and expected losses even greater than in the wild. Today, most good zoos routinely screen their animal collections for health problems. Thus, zoo animals tend to live longer than wild ones. Certainly, few people would argue that we should leave dangerous conditions untreated because it is unnatural to introduce medical care.

Closely related are the topics of adequate general husbandry, and safety of the maintenance staff. Having removed much of the natural opportunity for predation, keepers may become the one source of potential prey for the more lethal species. Efforts directed toward producing naturalistic appearances must not excessively reduce ease of danger-free access to the collection. While we may want to provide the tiger a spacious and naturalistic environment in which to display its beauty and romp happily, procedures and facilities must simultaneously be developed to provide for ease of restraint during routine servicing of the enclosure.

With many species, the very fact that they are in more severely limited confines produces special needs for unnatural sanitation procedures. Arboreal primates provide a simple illustration. In the wild, the problems of coprophagy (fecal consumption) are much reduced because the monkeys' feces may sift down through the trees and be naturally recycled. Except in times of famine, fecal consumption is relatively limited. In captive facilities, substrates are typically selected for ease of maintenance and cleaning. This is partly cosmetic; zoo managers have learned that the public does not like to see large quantities of excrement lying around. Somewhat paradoxically, these easily cleanable ground materials typically isolate feces in ways that make them more prominent for the inquisitive primate. Many of the best and most widely used primate foods are considerably less attractive than the stool in terms of texture. This combines with the fact that the majority of captive husbandry involves feeding according to a keeper schedule rather than at the animal's whim. Work is being accomplished on diet additives that give promise of reducing or eliminating problems of coprophagy, and active free-feeding environments also help to diminish this problem. In one study conducted by Ann Littlewood and myself, we found that ingestible substances which make the feces unpalatable are an effective means for sharply reducing coprophagy.

Besides their major missions in conservation and reproduction, the best zoos are largely educational institutions. One of the most exciting educational experiences in the lives of many children is the opportunity to behold the wondrous size of the elephant, height of the giraffe, speed of the cheetah, or flexible responsiveness of the dolphin. Yet, in the wild, it might take days of search to find opportunities to view some of the most attractive animals. For some species, it would be virtually impossible for the untrained observer to see them at all in nature. Consequently, public wild animal facilities strike another compromise. While they must reduce nature to the extent that they make residents visible to visitors, they also strive to produce naturalistic appearances. For some species, the use of one-way glass, artificial light, etc., may provide some psychological protection against the "intrusion" of visitors to their habitats. But for most zoo residents, careful periods of habituation to their captive home seems adequate. Thus, areas may be constructed to give some appreciation for a small portion of the species' native home, but reduce places of privacy to improve visitor opportunities for observation. An honest evaluation must conclude that we can work to provide microcosms of the animal's native home for education of the visitors, but that this does not mean we have produced a "natural" exhibit.

This leads to a final major point about the implausibility of natural zoos. It is expensive to produce highly naturalistic environments for many of the most commonly exhibited species. Most zoos work within severe financial constraints, and it takes careful engineering to ensure that new naturalistic exhibits do not introduce unbearable material and maintenance costs. In viewing one new naturalistic exhibit, I watched a zoo director slowly turn white with anguish as the freshly introduced tigers systematically destroyed young trees which had been purchased for several hundred dollars.

NATURAL IS NOT A SYNONYM FOR GOOD

The spirit of the times tends to make us champion popular ideas until they become virtually unquestioned. Besides the implausibility of naturalism, there are a number of reasons to suggest that the best interests of captive animals may not be served by making their state as "wild" as possible. Sacrilegious as it sounds, it is possible to improve on nature in ways that each of us has come to expect, if we were only honest about it.

Most obviously, unnecessary dangers can be avoided with carefully engineered habitats. In the simplest sense, even the most elaborate wildlife parks do not expose their more valuable species to the natural dangers of flood, famine, and predation. Most of us would probably agree that it is "better" for the animals not to be eliminated by these natural hazards. However, eliminating potential dangers carries with it the difficulty of eliminating a considerable proportion of the stimulation that nature ordinarily provides.

A related problem is the tendency for some observers to confuse *natural* with *traditional.* Frankly, many zoo workers see any change in routine as aberrant or unnatural. This is hard to justify if one looks at the native habitat of most captive species. In the wild, the rule is variety and change. Part of the beauty of animals lies in their adaptive responses to ever-changing climatic conditions, to appearance of new predatory dangers, and to problems of spatial distribution. Without radical change in exhibit design, we will continue to see only occasional fragments of stalking behavior, flight from predators, and other *species-typical* responses to natural stimulation.

PROVIDING FOR SPECIES-TYPICAL BEHAVIORS

The variety of potential methods to reintroduce excitement and opportunity for captive animals is limited only by the imagination of the designer. New exhibits should provide naturalistic ways for animals to obtain their own food. In some cases, this may literally involve the capture of live prey (see Chapters 5 and 11). Carl Cheney (1978) has proposed a number of such exhibits:

Education of the public, especially young people, is essential in order to explain that predation is the only way of life possible for some species. Development of a wildlife ethic is most easily accomplished with children, who have little vested interest in either total destruction or uncritical protection of animal individuals and species. It is important for a complete understanding of predation to realize that predator activities serve a positive, essential function in nature. A significant point is that man's intervention, in the form of prey-habitat perturbation, often unbalances the delicate predator-prey ecosystem, thus wreaking havoc and frequently forcing predators to prey upon domestic animals. Predators, existing at the top of the food chain, are very susceptible to ecological disruption. We are at a point in history where we can continue eliminating predators, and suffer the consequences of our actions, or we can exercise foresight and initiate some drastic modifications in our own behavior.

Many of Cheney's ideas require little in the way of added expense or care. Unquestionably, the real impediment to their implementation is the zoo administrator's worry about human revulsion at seeing the occurrence of death. For example, Cheney suggests that instead of feeding "freshly killed mice" to many species, one might present them still living. Raptors in general and owls in particular would be "naturals" for this kind of exhibit, and with careful, sensitive graphics, it should be possible to use this as a powerful educational tool. Some zoos have begun exploring the possibility of releasing rodents into raptor flight cages, but to date I know of none which do this routinely during public visiting hours.

For those species where it may be unpalatable to the public to allow live capture and feeding, or where it is otherwise impractical, the very least that can be done is to provide some animated prey (e.g., see "The Serval Project," Chapter 14). Although it is most desirable to work from the ground up on new exhibits to provide natural appearances, it is certainly possible to provide increased behavioral opportunities for zoo animals by using existing exhibits and limited budgets. The first such exhibit that we modified provided stimulation for gibbons *(Hylobates lar)* for more than six years and was initiated with surplus equipment and materials costing less than $200. The cage in which the exhibit was begun was a classical example of the modern mausoleum-prison-style zoo. Yet we were able to establish a simple paradigm that allowed the gibbons to exhibit their beautiful species-typical behavior of leaping and brachiating while obtaining food whenever they desired throughout the day. Later sections will describe a number of other examples illustrating that increased space usage and activity can be accomplished for widely varying species in existing exhibits that are far less than ideal.

PRELIMINARY PLANNING

The need for careful, extensive library and field research before modifying captive animal environments cannot be overemphasized. Besides helping to guarantee a naturalistic flavor to new behavioral paradigms, a thorough review of the animal's behavior may help to prevent calamities. For example, Hediger has emphasized that to encourage felines to jump on an apparatus placed in their cage may be deleterious if they can snag themselves and fall in awkward positions. In nature, they do not live in substrates as hard as those commonly found in zoos, and their anatomy has not been built to withstand the impact of falls onto these surfaces. Careful preliminary study of the species is bound to help suggest a myriad of ways to improve upon current exhibits and to design into new ones those things necessary to provide active environments.

CONCLUSION

Although the title of this chapter suggests that natural zoos simply do not exist, careful research and design can introduce species-appropriate behaviors to the zoo in much greater abundance than one sees in them now. The educational impact of an exhibit in which a visitor can watch a reptile moving to adjust temperature, a tiger pouncing upon prey—real or artificial—or an otter sliding into the water and searching for passing fish far exceeds narratives that require us to imagine the animal's behavior in its natural habitat. If we are to justify the removal of wonderful creatures from the wild, surely we must maximize their educational utility and provide species-appropriate opportunities in captivity.

3
THE FIRST PORTLAND PROJECT

Since most zoo animals have environments and husbandry procedures that do little to maximize species-typical behaviors, it was no challenge to find many alternative targets for the first project. We selected the white-handed gibbon *(Hylobates lar)* because the zoo's existing protocol seemed especially inappropriate for the species. In the wild, gibbons spend much of their time high in the trees of Southeast Asia. They are so ill-adapted for moving around on the ground that many observers have made reference to the gibbons' apparent dancing as they hold their hands up high, balancing themselves for bipedal movement. How strange that most zoos choose to throw food on the floor of their cage and have the gibbons eat in this markedly unnatural manner.

The white-handed gibbon has been the subject of a number of field studies (e.g.: Carpenter, 1940; Chivers, 1972). Most field workers have acknowledged the difficulties in completely describing familial and assortative behaviors because of the speed and generally private nature of the species. Social acquisition of behavioral skills, such as food collection, has been examined to a limited degree. In one case (Chivers, 1972) a comparative analysis has been made of the social behavior of the white-handed gibbon and the siamang *(Symphalangus syndactylus)* in the Malaysian Peninsula. Much of Chivers's data concerning the ecology and behavior of the gibbon is drawn from the work of Ellefson (1967).

In contrast with the siamang, which has not easily habituated to the presence of man, the white-handed gibbon continues to be found in most of the remaining forest areas of Malaysia. Most of the data suggest that gibbon territories are about 40 hectares in size, although there is considerable variability. Family groups occupying these territories are monogamous in structure, and typically include male, female and their young. Daily routine in the forest begins with "morning calls" given shortly after dawn as the gibbons leave their night positions and forage toward their boundary. Although the group may forage in separate trees much of the time, they occasionally find "preferred" trees in which they all come together to feed.

Since the first project was begun with no substantial budget, the development of many naturalistic opportunities was precluded. However, we knew that in Malaysia, the white-handed gibbon spends as much as 40 percent of its day in foraging and gathering food "on the run" (Ellefson, 1967). We could not provide fig trees for the gibbons to move among, but we were determined that we could give them the opportunity to eat up high and promote increased healthful species-typical activity in the form of leaping and brachiation.

Gibbons have been frequently studied in captive situations with special emphasis on their development (Rumbaugh, 1965) and capabilities such as brachiation (Andrews and Groves, 1976) and discrimination learning (Beck, 1967; Essock and Rumbaugh, 1978; Gossette, 1973). The acquisition of food-contingent behavioral skills by gibbons in captive situations has provided results that are often difficult to assess because of the apparent inappropriateness of task requirements (Beck, 1967; Essock and Rumbaugh, 1978) or motivational factors (Gossette, 1973; Harlow, Uehling, and Maslow, 1932). Results of these investigations in combination with field studies encouraged us to design equipment that would allow ample room for unrestricted movement by the gibbons.

When we began the work, the residents were Mama, an adult female, her two juvenile offspring, Harvey and Kahlil, and her infant, Squirt. None of us anticipated the rich results which these simple cage additions would produce for the next six years.

The gibbons' home in the Portland Zoo was furnished in traditional institutional fashion—concrete floor and walls with a cyclone fence on the front and top. Steel pipes had been inserted in a haphazard fashion to allow handholds for movements by the gibbons. From time to time, a rope was hung in the center of the cage to allow the residents to entertain themselves by swinging.

Detailed observations prior to the installation of the apparatus indicated that only one of the gibbons, Harvey, was doing much in the way of leaping or swinging. Occasionally he would show off for visitors, and when we came to make our regular observations, he would swing as many as seven times around a low bar in the cage, leap more than 15 feet semivertically, and slap the wall along the way before grabbing a handhold high in the cage (hence we monikered him Harvey Wallbanger). Typically, Mama, with Squirt hanging on, and Kahlil would simply sit and watch as Harvey went through his routine. Their major activity consisted of descending to the ground once a day, when food was thrown through a chute in the door, and competing for their favorite part of the ration. Once in a while, they would come down to the ground to stick their fingers in a trickle of water coming from the wall and scoop the water into their mouths or lick it off the back of their wrists in typical dainty gibbon fashion. Keepers would wash everything down the disposal in the early morning in an attempt to give a better image of what many visitors saw as an untidy cage with food laying around on the floor.

THE FIRST APPARATUS

Fortunately for our plans, the cage had two windows located more than halfway up the back wall. Two stations were designed for installation in these openings. Each had a stimulus light and a lever, and the second station also had a food-delivery chute. There was a narrow passageway in the center of the primate house that included some holding areas for apes and monkeys, and gave access to the windows where the gibbon apparatus was installed. We soon learned by talking to the keepers that traditional laboratory feeding apparatus would not suffice. First, universal feeders sufficient to hold an entire day's supply of bite-sized fruits and chow would have cost us $5000 which the budget did not provide. More importantly, any commercial apparatus installed in the narrow passageway would have made the keepers' job so difficult that the work would have been immediately unwelcome. So, we designed a simple food belt by using donated mylar film laced into an endless loop. This was driven by a motor scrapped out of a coffee machine and some programming apparatus that we made from surplus parts (see Chapter 13 for construction details). Necessity can be the mother of invention and, as things turned out, this design for a feeder proved excellent for use throughout the Portland Zoo as well as in projects accomplished in other places. Occasional replacement of worn-out film conveyor belts or motors have been the limits of our service needs.

Figures 3-1a through 3-1d illustrate the final general sequence that we asked the gibbons to accomplish. They learned that when the first light went on, a response to that lever would activate the second station (approximately eight meters away). Whether the second response was made by the same animal who initiated the work or by another animal was up to the gibbons themselves. Indeed, this choice provided us with some of our most interesting data. Initially, we taught Mama, Harvey, and Kahlil that a response at the payoff station when the

Figure 3-1a. Gibbon *(Hylobates lar)* responds at first station.

Figure 3-1b. Gibbon leaps after initial response.

light was on would give them a piece of food. This stage was easily learned by all three animals in a matter of a few days. Next, we progressed to the much more difficult step, encouraging them to move from the place where food was earned across the cage towards station 1. The gibbons were such adept students that in a matter of weeks all could accomplish the complete sequence. We introduced an intertrial time of two minutes to spread out the food earning and to reduce the probability that one or more of the animals would simply sit at the payoff station.

SOME EARLY RESULTS

Harvey, whom you will remember had always been the most active animal in the group, became so adept that he could complete the en-

Figure 3-1c. A single, brief grasp in swinging to feeding station.

tire sequence in less than two seconds. (In several sections of this book, I will refer to specific animals with personal pronouns such as "whom" because, despite its literary accuracy, I have always found "which" and "it" to be cold-sounding terms for animals I knew well.) After a time, Mama and Kahlil would only occasionally complete sequences themselves. Most of the time they seemed satisfied to let Harvey do the work while they sat at the payoff station and collected the fruits of his efforts. Harvey's behavior dramatically changed after a short period of this exploitation. Much to our surprise, his behavior was markedly different depending upon which gibbon was in a position to take the food after Harvey had helped to earn it. He continued to work when his mother got the payoffs, but developed a whole new style when Kahlil was in the ready position.

Figure 3-1d. Eating a piece of fruit actively earned. *(Sequence of gibbon photos by B. McCabe)*

Since the gibbons had unlimited time to respond after the first stimulus light went on, there was no greater premium for immediate movement to station 1. Further, if they did not use up their day's supply of food, free food was given at night anyway. (After initial training, the gibbons almost always chose to earn their food rather than to have it free.) So, when Kahlil would approach the payoff station, Harvey would not respond at station 1. Instead, he would repeatedly wait until Kahlil moved at least halfway across the cage before beginning to work. Many spectacular races ensued. It was unpredictable who would get the food, but surprisingly there was no fighting. This mode of behavior lasted for a long time, and we decided to interface the public with this exhibit in a way that would make the intervals more random.

The idea of allowing visitors some means to interact with the animals has almost always had appeal for zoo managers. Unfortunately, the majority of traditional interactions tend to be degrading or otherwise disadvantageous for the residents. Excessive food thrown in by the public, encouragement of begging behaviors, and protocol that leads visitors to think of the zoo residents as bizarre, strange, or worthy of mocking are still unfortunate parts of many zoos. We wanted to develop a system whereby the public could interact to help the gibbons obtain good food in measured quantities, and through which they could learn to appreciate the capabilities and behavioral beauty of these animals. At first, it was planned that a button would be installed that allowed the public to shorten the intertrial interval. But, with the urging of one of our students who needed an oscilloscope to continue some research, it was somewhat reluctantly decided to have the public donate to the development of further animal enrichment procedures. A coin box was installed with the following legend:

Research Contribution:
Ten cents will start a trial when the light on the box is lit. The counter shows the total number of pieces of food earned by the gibbons today.
Animals are not machines and the gibbons may choose not to respond when the light is turned on. All money collected here will be used to develop more activities for our animals.

It is important to note that even after the installation of this new apparatus, the maximum time that the gibbons ever had to wait from their last piece of food to their next was two minutes. If no one came along or no one decided to contribute, the light always went on in two minutes anyway. We thought we might raise a few hundred dollars for the oscilloscope and other equipment to begin some of the projects covered in later chapters. Instead, the public donated more than $3000 in dimes in the first year, showed great interest in this method of contribution, and wrote hundreds of letters indicating their hope that the work would be extended to other species.

As time progressed, zoo management decisions and husbandry requirements led to changes in the cage population. Perhaps the most interesting outcomes of these changes had to do with the way the gibbons acquired responses with no further training from the staff.

SQUIRT GROWS UP

The noisy milieu in which this work was carried out made it impossible to come to precise conclusions about the method of social transmission of learning. But, it *is* very clear that the gibbons were able to learn from each other. Our first evidence came as the infant matured. As Squirt was weaned and increased his movement away from Mama, he eventually began to make complete chains of responses. These were interspersed with all sorts of play and species-typical teasing and mock fighting between Squirt and the adolescent males. Eventually, when Kahlil was moved to another zoo, Squirt began to take over a substantial portion of the food-earning for the cage, finally coming to be almost as speedy and spectacular as Harvey. Thus, a gibbon whose only opportunities for interaction during the training periods amounted to clinging to his mother as she earned food learned not only to feed himself in this active paradigm, but he also provided a considerable amount of food for the other gibbons and apparently enjoyed the activity.

At one point in our work, it was decided to replace some electromechanical programming and recording equipment with new solid state devices. This gave us an unexpected opportunity to evaluate the gibbons' interest in actively earning their food. It was decided that since they had long been able to have bite-sized food high in the cage rather than having rations thrown in on the ground, we would continue to provide precut food throughout the day by passing it through the apparatus windows during the time the equipment was disabled. The gibbons would take food in this fashion, but appeared remarkably annoyed at the lack of responsiveness of the apparatus. For example, we repeatedly saw Mama and Harvey with their hands full of food moving to the stimulus lights and levers and trying to get them to respond.

It is difficult to think of this as some sort of compulsive behavior because for years the gibbons' responses to the first station had been primarily limited to periods when the stimulus light was on. Now they would move to it, apparently imploring it to turn on and respond to their efforts, even though there was free and abundant food available, identical to that which they earned. As suggested in the introductory chapter, we believe that this is another bit of evidence that animals prefer responsive environments and, as long as the demands are not too great, they prefer to work for food rather than having it placed at their feet.

THE COURTSHIP OF HARVEY AND VENUS

About two years after the establishment of the gibbon project, another young adult male lar was donated to the zoo. The staff decided to search for a mate for this handsome fellow whose name was Milo.

Milo had an interesting history, having lived most of his first seven years with the nice woman who had donated him to the zoo. Consequently, Milo knew much more of people than he did of gibbons. No one knew exactly what to expect when he was introduced to the lovely dark-colored mate who was acquired from the Oklahoma City Zoo. Because of the novelty of introducing an animal who had been wild caught and reared with gibbons to one who had been reared by a human, the media were invited by the zoo's public relations staff to witness the introduction.

The first day's meeting did produce some interesting footage for the local TV stations. One of the announcers named Milo's new acquaintance "Venus." The window entrance halfway up the cage wall was opened, and Venus proceeded to chase Milo all over the cage. Eventually, she cornered him. Milo had no choice but to come into contact with her and immediately began to sniff around. At that point, Venus turned and coquettishly ran away with Milo in hot pursuit. All of this was shown on the evening news to the music of the currently popular tune, "Games People Play."

Despite the great promise of this initial encounter, Venus and Milo never really hit it off as mates. They learned to solve the discrimination problem which will be touched upon in a later chapter, but most of the time Venus dominated Milo and warded off his advances toward her.

For a couple of years, these two gibbons were left together in a cage identical in size and general conformation to the one described above, but without the enrichment devices. During all but a few months of this time, they had access to a small panel in their ground level door which was used in the discrimination studies. This was the sum total of Venus's formal education. Subsequently, a management decision was made to attempt mating Venus with Harvey in the cage where the original apparatus was available.

We documented the first eight hours of the introduction on videotape and then began to monitor the behavior in this cage with our routine procedures. This included two hours of daily observations plus constant automatic monitoring of the responses to the levers.

Upon his introduction to Venus, Harvey began one of the most spectacular displays we had seen from him or from any other gibbon. He leaped and brachiated all around the cage, made several food-earning passes and generally seemed to be entertaining Venus. Venus, whom we had always seen as a rather aggressive female, not only was gentle with Harvey, but she would also actually move out of his way in order to make his performance easier. Within the first hour, our videotapes show Harvey approaching Venus and kissing her on the cheek. Harvey provided much of the food for the first several weeks, with Venus watching intently and following along on occasion. Within a matter of days, Venus had learned to obtain food by making complete sequences between the stations, and she greatly increased her productivity until there were many periods in which she earned most of the food. These gibbons were observed mating, and there was hope that they would develop a family.

CONCLUSION

Despite its excessive simplicity, this first project proved to be of significant value from a number of standpoints. A printout counter was installed to accumulate data in two-minute bins for almost a year. This allowed us to make some assessments of seasonal variables in feeding behavior in this captive situation; the success from a visitor's standpoint encouraged the development of many other projects in Portland and other zoos and we were able to generate some interesting ideas about social acquisition of learning. But nothing was more exciting to us than the knowledge that we had given some behavioral freedom to the gibbons in the sense that they could earn their food when they wished rather than being dependent on the whims and timing of the staff.

It was not a particularly attractive exhibit, with its austere globes and levers and its concrete and cyclone-fenced construction. Nevertheless, this first project did demonstrate that even in this highly artificial situation, species-typical behaviors could be encouraged and animals could learn, and apparently preferred, to earn their own food.

4

A TOKEN ECONOMY

A factor which we have not dealt with in describing the early gibbon work is the matter of quantification of sharing and stealing behaviors. Frankly, it was virtually impossible to find an objective method for identifying those cases where gibbons willingly gave up their earnings to others versus the times when they were stolen. In planning our second venture to increase space usage and species-appropriate activities, we decided to use a token economy in the hope that it would help us to quantify altruism and sharing versus stealing. The group selected for this work was the beautiful West African diana monkey *(Cercopithecus diana)* family which lived in a cage of construction similar to that of the gibbons' home. When we began, there was Beulah, a 16-year-old female, her mate, Rocky, who was eight years old, an adolescent male (Butch), and an infant male (Kid).

The program began with an effort to teach the dianas to exchange tokens for food (Figure 4-1). At first, we installed a slot with ample space for the deposit of tokens and balanced the tokens in this slot, allowing the monkeys to push them through. It was hoped that by successive approximation, we would be able to get the monkeys to pick up their own tokens and drop them into the slot. Care had been taken to make the coin of the realm difficult to fit in the animal's cheek pouch and impossible for them to break or swallow. One of the local electronics firms was kind enough to donate special nontoxic materials which they poured into appropriate dies to make tokens whenever they purged their molding machines.

In spite of this nice community cooperation and what we thought was good planning, the first step was a dismal failure . . . Unlike some

Figure 4-1. Experimenter hands diana monkey *(Cercopithecus diana)* a token which can be exchanged for food. *(Photo by H. Markowitz)*

of the apes which are so facile in manipulating objects such as tokens, the dianas did not seem to be able to consistently orient tokens vertically enough to get them through the slot. Since the major interest was not in response topography, but rather in development of a medium of exchange for the monkeys, we went back to the drawing board and installed the V-shaped slot shown in Figure 4-2. This slot funneled the tokens past a hidden photocell which triggered off the mechanism that delivered food to the monkeys. Within one day of installation of the new apparatus, two of the dianas learned to deposit tokens.

Several months were spent trying to ensure that each of the monkeys would learn to spend tokens. Alas, no kind of training that any-

Figure 4-2. Diana monkey inserts token to obtain fruit. *(Photo by B. McCabe)*

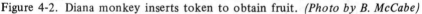

one could devise was successful in getting Beulah to deposit them. She would drop tokens alongside the slot or throw them at the wall, but never managed to insert them on her own. Finally we decided to progress to the next stage since the youngsters had learned well and so had Rocky. It turned out that this was probably a fortunate decision since several years later, after a number of offspring had been born and learned to deposit tokens by watching the other dianas, Beulah was still not effective in spending tokens.

When the value of the tokens in exchange for favorite foods had been established, we installed a station four meters up the back wall and diagonally opposite the automat (Figure 4-3). This station included two stimulus lights and a chain that dangled from a lever. When

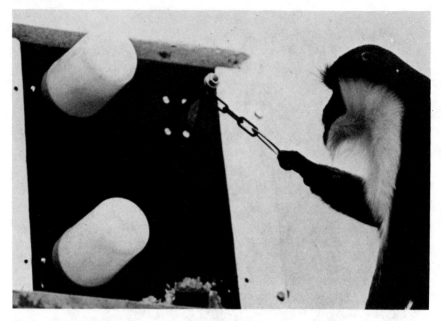

Figure 4-3. Diana monkey pulls chain to obtain token. *(Photo by B. McCabe)*

the bottom globe was illuminated and the dianas pulled the chain, a token was automatically delivered to the monkeys through an opening in the apparatus. Once again, the training went smoothly for all of the animals except Beulah. Some marked differences began to emerge in the way Rocky reacted to the apparatus as compared with the juvenile Butch, who was also quite adept at earning and spending tokens.

In general, Rocky would take over at the beginning of the waking day and would begin collecting and exchanging tokens for food until he was relatively satiated. During this time, each of the young dianas attempted to sit on the platform and share in his earnings as he deposited the token. He would unceremoniously "wipe" them away from the automat with a single deft arm swing. But to everyone's astonishment, his behavior was categorically different with respect to his mate. On a number of occasions, I actually saw him take two tokens, drop one in for food for Beulah and one for food for himself.

There also was some tendency at first for Rocky to hoard tokens. On one recorded observation, he was seen to have three clenched between his teeth and four in a rear paw. However, this behavior rapidly faded since, in this as with other designs, there was no great compulsion for the dianas to earn food. They were deprived no more than any other zoo animals, but instead had opportunities to earn their food earlier in the day if they chose to, which they always did. Before too long, in the middle of the afternoon earned tokens could be seen laying around for any monkey to spend.

Even though Beulah had not mastered either part of the paradigm, once again it was arbitrarily decided to progress to the next step. The final installation is shown in Figure 4-4. A long rope-like chain was suspended along with a signal light from the top of the cage near a matrix of bars. It took virtually no training to get the animals to respond at this level when the light was on because the youngsters enjoyed swinging on the chain from the moment it was installed. In order to call the monkeys' attention to the availability of this station, the top stimulus light on the token delivery station was used.

To review the complete sequence: no more than two minutes after the last earned piece of food, a light went on at the top of the cage and simultaneously on the panel where tokens were delivered. When any of the dianas swung on or pulled the top chain, this turned on the second light, and a response midway up the back of the cage delivered a token. These tokens could then be spent any time that the animals chose. They could accumulate them, they could spend them one by one as earned, or give them away (a lot of bad jokes have been made about the dianas going out and buying themselves a monkey). The dianas quickly learned the entire sequence and eventually a coin box similar to the one used for the gibbons was installed to allow the public to serve as a random interval generator and a source of research donations.

Activities generated in this family of diana monkeys were even more varied than those unusual work conventions described for the gibbons in Chapter 3. Visiting primate experts regularly expressed the feeling that they had never seen more active or "alive" monkeys in captivity. Many also described the interesting parallels between behaviors which could be seen in this artificial milieu and those that they had observed in the field. Here was a tool with which we could look at the development of dynamic relationships in the family, the ways in which work was apportioned, and the relative success of family members in token and food acquisition.

Figure 4-4. Completed diana monkey installation. Male responds at top station while female collects token. *(Photo by N. Lee)*

Rocky was very good at earning tokens and spending them for food, but his efficiency was somewhat reduced by the repetitive nature of his behavior. For many months, we observed that he always approached the top chain in the same direction. No matter where he was in the cage when he decided to respond, he always moved to the right-hand wall and always approached the first response in a direction which required a right-hand turn. Then he would turn around, go back to the same point in the cage, make a left-hand turn, and work his way systematically down the wall until he got to the token delivery station.

Butch was much more flexible. He quickly learned to do little loops around the bar from which he reached to pull the top chain, and swung directly to the second station in less than a second. One marvelous day, I watched Rocky standing back and observing Butch for a long period of time as his offspring earned token after token, taking shortcuts along the way. Finally, Rocky went over, made his usual first response, but then took a little loop shortcut to the second station. Anyone who has watched the same group of monkeys day after day for years may not think me too foolish if I tell you that he had the most self-satisfied look that I could conceive as he held up the token from this trial. For several days, he would intersperse these shortcut trials with his standard procedure. Eventually his responses were much more varied, but he never became quite as flexible as his youngsters.

Beulah's responses were increasingly efficient, but not in the sense that we had planned. On rare occasions, she made a solitary response at the top of the cage, but most of her time was spent at the token delivery station. Here she watched to see when one of her youngsters swung on the top chain. Then she would either wait for them to come down and earn the token directly, or occasionally she would pull the chain and have the token waiting for them. When she felt like eating, Beulah would encourage the youngsters to go spend the tokens, sometimes patting them on the rumps and occasionally just shaking her head in the direction of the food exchange apparatus. After the younger dianas deposited the tokens, she frequently took the food from them.

None of this was terribly disturbing from a husbandry standpoint since exactly the same sort of thing had been seen with ad-lib food. When piles of food were placed in the monkeys' cage, Beulah frequently snatched the more preferred morsels from her offspring. This brings focus to a critical factor in innovative behavioral work with animals. All of the special behaviors typical of primates may be seen by casual observers as a result of the new protocol. I recall overhearing one visitor, among many who were happily watching the active dianas, who was disturbed by Beulah's "stealing" behavior. Fortunately the remedy was easy in this case. I simply called her attention to the adjacent cage where the monkeys had their food thrown on the floor. As the youngsters reached for preferred pieces, their parents would grab them away from them.

But in this enriched situation where the animals could earn food at their leisure, they were no longer entirely passive about giving away their fruit and chow. Instead, they learned many clever ways to reduce this exploitation. For example, Butch would take the token to the slot, fake putting it in, and then run off and play for a while. He would watch his mother from time to time while playing and return occasionally to click the token against the slot. At first, Beulah would stand expectantly waiting for the food delivery, but eventually she involved herself in some other activities. Then on some random trial, when Beulah was not hovering nearby, Butch would spend the token and thus have the food himself.

No one has come up with a really adequate explanation of Beulah's behavior. She certainly had ample time to learn the paradigm, she did occasionally make responses, and she had no apparent physical anomalies. Her eyesight was keen enough to catch flies, and her tactical capabilities were very apparent in her stealing behavior. Once during the time that we were struggling to teach Beulah how to spend her own tokens, one of my good friends suggested that she might be the object of male oppression. I proposed that we make a pilot study to see whether or not the suggestion had any credibility.

My friend watched for 12 hours and then reported her results with a kind of funny grin. Rocky had not taken any of Beulah's tokens until one critical trial about half-way through the observations. On this trial, he sat back on a bar about six feet from the food station and watched Beulah repeatedly drop a token alongside the chute. Rocky began to rock back and forth in a rather impatient manner. Finally, he hopped over beside her, picked up the token which she had just dropped, put it in the slot, and let Beulah have the food. The same thing happened twice more during the brief study period, and the observers concluded that if there was any oppression, it was of the mental variety.

These admittedly anthropomorphic descriptions of the animals' anecdotal behaviors may give the reader some hint of the highly individualized styles of the monkeys. We certainly did not succeed in answering the question about altruistic versus stealing behaviors in any defensible systematic way. The sharing or stealing of food was simply displaced one more step to the point where the tokens were exchanged for food. But, the elusive behaviors by the youngsters, the deferential behaviors by Rocky, and Beulah's encouraging and food-snatching behaviors did provide strategy for more adequate guesses.

In general, it seemed clear that early in the waking day, when the animals were hungrier, they were both more acquisitive and tighter with their earnings. Beulah was also more aggressive in trying to snatch food from the paws of the younger monkeys who were very apparently reluctant to give it up. After an hour or two of food-earning, things got much looser. Often tokens were earned apparently for the fun of the exercise and were essentially available ad lib for whoever wanted to spend them. The monkeys remained vigorous, the adults were re-productive, and the youngsters went through all of the usual play-fighting bouts that one sees with healthy, fit animals. The work load shifted from time to time, depending upon the stage of Beulah's preg-nancy, the number of offspring in the cage, and temperature variables. However, the dianas continued to prefer to earn almost all of their food, and observers continued to marvel at their abilities.

For a number of years, we had the daily pleasure of visiting the dianas, usually the most active monkeys in the zoo. Several students accomplished complete research studies by simply selecting specific behaviors for focus while the monkeys arranged the details (e.g., Bandura, 1974; Soper, 1974). We eventually replaced some of the nylon chains with more naturalistic rope, and while I always dreamed of real trees for these monkeys to scamper in, it was rewarding to know that they had more chance to arrange their own schedule of food-gathering activities than do most zoo animals.

SOME FINAL THOUGHTS AND FUTURE PLANS
FOR DIANA MONKEY RESEARCH

Although dianas are among the most colorful, active, and inquisitive of the monkeys exhibited in zoos, their behaviors in the wild and in captivity have not been extensively studied until recently (Morike, 1973; Markowitz, 1978). Consequently, many of the assumptions about husbandry requirements, health care delivery, and social dynamics have been extrapolated from information about other cercopithecines (e.g., Struhsaker, 1967; Rowell, 1971).

In the wild, these frugivorous West African primates occupy the middle and upper stories of their forest habitats. The first behavioral observations planned for our next diana study are largely designed to evaluate the expectation that these arboreal animals will greatly benefit from the opportunity to utilize upper tree branches for species-typical locomotor patterns and behavioral displays. As Meyer-Holzapfel (1968) has stated, "Many traditional sterile cage designs severely limit behavioral possibilities and contribute to stereotypical activities by the inhabitants." Providing increased vertical space and environmental complexity should increase diversity in activity profiles, and may potentiate positive aspects of the natural social dynamics in this integrated group.

Much too often, expected effects such as we have described above are taken for granted rather than documented in detail. This has two deleterious effects: the apparent lushness of the new home may lull us into ignoring potentially harmful effects on the residents, and necessary readjustments of the environment and husbandry procedures which should be expected as a matter of course with a new exhibit may not receive prompt attention. In order to identify the problem areas as quickly as possible, intensive observation of the effects of increasing space and including abundant vegetation will be integrated with a program that tests new construction techniques at the San Francisco Zoo.

The existing diana monkey exhibit in San Francisco presents an ideal opportunity to provide a naturalistic environment for a well-established primate group while employing the newest techniques and materials available in zoological park design. Tiny exhibit cages are paradoxically surrounded by a lush miniforest currently totally inaccessible to the monkeys. Our plan is to demolish much of the concrete and steel and open the wooded area to the animals. Zoo visitors will have limited access to the dianas' home by means of enclosed pathways in the forest.

We have consulted with zoos and wild animal parks to establish the impediments to incorporating more natural materials in an attractive and economical fashion into the habitats of small primates. Primary among the difficulties are the rigidity and cost of traditional caging materials such as cyclone fencing and welded wire, and the fact that existing modular approaches to habitat improvement produce excessively stylized environments.

Significant departures from traditional exhibits such as the Brookfield Zoo's Tropic World and the primate areas in San Diego Zoo's Wild Animal Park are exciting and innovative, but many zoos are unable to undertake the construction of projects of this scope that currently require seven-figure budgets. Consequently, long overdue improvements and renovations are often left unaccomplished because of the hope that, someday, funding will come to fulfill idealistic master plans. The first aspect of our research will be to evaluate the use of new, economical materials in providing larger-scale naturalistic environments for monkeys, while allowing improvements to be accomplished in a modular fashion. A primary goal of this research is to establish methods allowing the richest possible life for the resident species while still integrating the themes and materials within the broad scope of primate habitats that a particular zoo may select.

The closest precedents for our work are encompassed in the lemur exhibit at the San Diego Zoo, the South American rain forest exhibit at the Panewa Zoo in Hilo, Hawaii, and the new arboreal primates exhibit at the Royal Melbourne Zoological Gardens. The San Diego exhibit's technique is similar to our plans in the sense that it incorporates trees and other botanical items and uses low-maintenance material (vinyl-dipped welded wire in one-half by one-inch squares). Our experimental high-strength polyethylene materials will provide several advantages: they can be woven into the trees, thus maintaining a natural skyline effect, they are flexible and resilient, providing a less rigid appearance, and they will allow us to take maximum advantage of existing trees and landscaping because of free-form capabilities.

Both the Panaewa rain forest exhibit and the Melbourne arboreal primate exhibit include the feature of a walk-through for the visitors. This helps to provide an impression of visiting the monkeys' home rather than viewing it from outside a cage. Our strategy is to combine this quality with a more complete feeling of actually entering the forest. This will be accomplished by using hardy existing trees rather than trying to grow new specimens that usually take considerable time for adequate establishment.

The small current cages (see Figure 4-5) include concrete slabs for the animals' ground surface. Although these slabs will be eliminated in the revised habitat, the renovated houses will be clad in wood and remain as available shelters, service areas, and temporary isolation units. Preliminary observations will be made from a number of standard locations both above and below the existing structures, but outside the perimeter which will be incorporated in the new exhibit. By · this method we hope to produce data which can be directly compared for the same residents under the old and new environmental conditions.

A smaller portion of the new exhibit will provide residence for talapoin monkeys *(Cercopithecus talapoin)*. This will allow an inter-

Figure 4-5. San Francisco Zoo—small concrete cages for diana monkeys. *(Photo by W. Wong)*

esting basis for comparing behaviors of two West African monkey species within the same forest grove. Even though they will be visually rather than directly mixed, there will undoubtedly be quantifiable changes in the dianas' behaviors as a function of the proximity of the talapoins.

An especially rich opportunity to evaluate the relative difficulty of introducing new animals to the existing groups is predictable for these particular inhabitants. The talapoin has a rather well-documented set of behaviors upon introduction of newcomers (Scruton and Herbert, 1972). When strangers are introduced, residents of the same sex attack a newcomer, and there is some tendency for the adult females to be more aggressive than adult males (Wolfheim and Rowell, 1972). As with most monkey species, juvenile talapoins and juvenile males in particular tend to be more active, innovative, and playful than mature animals. We have documented similar effects in the captive group of diana monkeys in Portland, and have also observed attempts at integrating new animals to several diana collections. Unlike the talapoins, in the case of dianas, all of the residents, but particularly the juvenile animals, tend to attack newcomers. The effects do not seem to be sex-specific.

Because of the enlarged environment and increased opportunity for flight and gradual integration with the existing group, the planned enclosure should provide opportunity for reasonably safe introduction of small numbers of new animals. During these critical periods we plan to have a minimum of three trained observers collecting data for at least 72 consecutive hours, or until behavioral dynamics become stable.

A mundane sounding, but important part of the research, will involve studying effective maintenance procedures for this naturalistic habitat. While the large domain and the comparatively small size of these monkeys means that there should be little problem in naturally recycling waste materials, problems of cleaning out unused foods must be addressed. Our strategy in studying the best methods to cope with this difficulty will begin with the establishment of several automatic feeding stations where the monkeys can obtain food on a unit basis throughout the day.

The establishment of regular feeding sites will also allow us to install automatic electronic weighing devices with which to collect data and help monitor animal growth. Besides providing data of value in animal care, these electronic readouts will be used to allow visitors opportunities to directly learn about the weight and location of individual animals. The electronic scales will also serve as "signature" devices to help new research trainees in reliably identifying specific monkeys. Automatically accrued weight data will be attached to each observational set for computer analysis. This will provide a point of departure in collecting reliable and easily measurable physical health data to correlate with behavioral findings. The development of routine preventive medicine criteria and techniques has been a long-term goal of our research (see Chapter 14).

Simultaneously with our studies of the behavior of the monkeys, we will assess changes in visitors' behaviors and attitudes plus time spent at the new habitat compared to the old. The fact that the public will have the essential impression of entering the monkeys' forest area allows the possibility for researchers to inobtrusively collect data from observation stations in the exhibit. This portion of the research will receive systematic analysis similar to that which we plan for animal behavior and environmental materials evaluations.

We hope by means of our study of visitor behaviors to evaluate the educational and recreational value of this exhibit and to help in evolving dynamic graphic materials that will maximize this impact.

5
PLANNING SOME NATURALISTIC
OPPORTUNITIES FOR
RIVER OTTERS

The sliding of otters, popularized in some television nature programs and Walt Disney productions, is not a frequently observable behavior in zoos (Myers, 1978). Yet a large number of zoos and aquariums have incorporated slides for their otters, and occasionally programs using reinforcement techniques have illustrated that otters can be encouraged to slide on a regular basis. These programs, including ones in which I have been involved (Markowitz, 1978), have always seemed unsatisfactory to me from both an educational and a zoological standpoint. The chance to watch these active and clever animals slide down a pastel swimming pool slide does little to give visitors information about species-typical behaviors. Another element that is missing from most exhibits is the beauty of the capture of prey by the otter. Fish, ZuPreem, and carrots from the refrigerator do little to stimulate activity in the otter unless they are deprived. This kind of food also eliminates the zoogoer's chance to observe some exciting predator-prey interactions.

For the past three years, some of my students and I have been doing the library and field research prerequisite to designing a naturalistic environment for the North American river otter *(Lutra canadensis)* (Figure 5-1). In this work, we decided, with the encouragement of some zoo administrators, to take on the difficult task of educating the public about the advantages of letting the residents capture live prey over humans providing prekilled food for them. This work may serve to illustrate an important aspect of behavioral enrichment in

Figure 5-1. North American river otter *(Lutra canadensis)* sits in barren exhibit. *(Photo courtesy San Francisco Zoological Society)*

the zoo: it provides unique opportunities for focused interdisciplinary research.

Recognizing the realistic limits of the San Francisco Zoo budget, we decided the first step was to begin our own active fish factory. Because of their hardiness, palatability, and prolific reproduction, we selected the African mouth brooder *(Tilapia mossambica)*. Several students are currently working on this breeding program, and this will provide an example of how preparing for a single enriched exhibit may generate many potential student projects and thesis materials.

One concern about feeding live fish to animals is the possibility that they will be more contaminated than those that have gone through the freezer. Barbara Biro is designing a parametric study to analyze the comparative parasitic and bacterial loads of the *Tilapia* which we raise under careful control versus that of fish taken from the regular frozen supply at the zoo. This turns out to be a very interesting and complex study that has brought together faculty members with expertise in parasitology, bacteriology, nutrition, and behavior to help Barbara with her work.

A project involving several students focuses on methods for obtaining maximum reproduction of *Tilapia*. We are planning breeding areas with separate light and temperature control in order to help to guarantee that there will always be some reproductive fish. Another study will use hormonal agents in the water that have been shown to affect development of *Tilapia* so that all offspring are male. Since, in this species, sexual determination causes a stunting of growth of the females, this would provide larger prey for essentially the same cost. It may be interesting to note that *Tilapia* have been cultured for food for a very long time. According to Hickling (1963), "The earliest known representation of a fish-culture in history—a bas-relief from an Egyptian tomb dating from before 2000 B.C.—shows a pair of small fish that can be identified as *Tilapia nilotica,* a species still abundant in the Nile valley." This aspect of the work, which started out as a simple search for abundant prey, has developed into a consuming research interest for many of us.

Although the pool where the work will be accomplished has concrete walls and a uniformly circular perimeter (Figure 5-2), we believe

Figure 5-2. Existing otter pool of San Francisco Zoo. *(Photo by W. Wong)*

that a face-lift can do a great deal to provide a more naturalistic appearance. Using tough but resilient materials, a surface which allows cleaning essential to zoo husbandry but less artificial-looking contours and terrain will be established. From a central island, there will be four areas that slope into the water at angles approximating those reported for otter entries in the wild. One common way in which otter slides occur at stream and river banks will serve as a model for this part of the development.

River otters often find convenient places of entry and use them repeatedly. Since they occasionally defecate when entering the water, their fecal matter, which contains a large proportion of fish scales, makes for a nice slippery route to the water. We reasoned that no health inspector or city plumbing agent would condone the exact reproduction of this natural edifice, so we substituted materials that become extremely slippery when wet, and we plan to use a continuous recirculating water source to moisten the slide area.

On one side of the pool, a waterfall will conceal a dispenser of live fish. Many hours have been spent discussing the merits from a recreational, educational, and species-appropriate standpoint of different methods of dispensing the fish. While at first consideration it might seem more naturalistic and logical to always have some *Tilapia* swimming around, we know that the city water and widely varying temperatures would lead to the death of many of the fish before they were consumed by otters. We also want to ensure that other necessary components of the otter diet are eaten, so an infinite or uncontrolled supply of fish would be unwise. Allowing the release of a small number of fish on a random basis is precluded for some of the same reasons. There is no way to ensure that the otters will choose to swim and hunt for food at the particular time of day when the prey are at their liveliest. Consequently, we decided to allow the otters' use of the slides to trigger the release of fish on a random basis. Programming controls will be microprocessor-based and allow easy schedule changes so that the proportion of fish released may be adjusted on an empirical basis. This will also provide flexibility to try two major alternatives: adjusting maximum prey frequency to the zoogoers' schedule or to the schedule dictated by the otters.

In final form, the rationale is rather straightforward. Otters will find that occasionally when they slide into the water, fish swim by.

We believe that this will encourage the use of the slides as entries for many of the same reasons that otters go back to good fishing places in nature. Visitors will unquestionably find more activity and more that is instructive about the nature of the otter by watching it pursue live prey. The university will have the privilege of having its students work with these animals in the zoo and in return will help to develop unique educational and recreational opportunities for zoo-goers. We make no pretense that this will bring nature to the zoo, but it will certainly bring more naturalistic opportunities to the otters.

There are a number of reasons why we selected fish as the first live prey species to be used in our behavioral work in zoos. The first and most evident reason is pragmatic: experience with zoo visitors suggests that a greater proportion oppose viewing predation when mammals are consumed. An interesting thought in this area is that most of the butchered mammalian flesh that we ourselves consume bears little resemblance to the intact creature. On the other hand, whole fish often appear in the packages and cans which we bring back from the market. Perhaps most kids grow up recognizing that they're eating the "real thing" when it comes to fish.

A second advantage in using these fish is that they do not require great expanses for mobility, nor do they characteristically exhibit the freezing responses which many other prey might show when trapped by the predator. Consequently, the public will be able to study active capture behaviors rather than simple exposure of live prey in excessively narrow confinement.

Teaching our children and each other that some animals die in order for others to go on living will not reduce reverence for life. Instead, a clear understanding of this fact may reintroduce some of the appreciation of that segment of nature that commercial food preparation removes from most of our conscious attention.

6

AN ALTERNATIVE TO
BEGGING BEARS

Bears have traditionally been one of the favorites with zoo visitors—all too often for reasons that can at best be said to be demeaning to the animals. A visit to an average zoo will almost certainly provide the opportunity to watch visitors encouraging bears to beg and then throwing them ill-chosen food as a reinforcement. It is common knowledge that many zoos have the "Monday morning syndrome" in which animals fed unmeasured amounts of food, such as marshmallows, by large weekend crowds, become diarrhetic and reluctant to eat their zoo diets. Why does this sad and inhumane procedure go on in an age when increasing numbers of zoo professionals and people in general are beginning to attend to animals' rights and needs?

The answers are direct if not simple or defensible. First, any honest person must admit to the positive feeling which comes from any effective interaction with these wonderful animals. Much too frequently, the only chance to interact comes by means of reinforcing unusual behaviors such as begging. In a number of cases where zoo directors have established "no public feeding" policies, there have been overwhelming public outcries because people feel disenfranchised from traditional interactions with "their bears." Fortunately, some administrators have the stamina and community support necessary to stand their ground on this issue. However, it is understandable that there is some reluctance for institutions which always seem to be in financial or political jeopardy to do anything which disquiets their clientele and their fiscal base.

Another rationale for continuing old policies is, again, the very fact that they are traditional. Bears, seals, and a myriad of other animals learn to perform remarkable acts to receive preferential treatment in public feeding. Because of the variety of routines that are sometimes evident, one might conclude that this was a cute and largely voluntary behavior. A more careful analysis indicates that these behaviors are unwittingly shaped by visitors. If an animal commits a particular act that increases the probability that it will receive a preferred food, this in turn increases the probability that the same act will be repeated and ultimately honed to a point where it becomes an idiosyncratic skill.

One might ask how the first of these behaviors is initiated by the animal, if not voluntarily. The most parsimonious answer is clear on the basis of abundant laboratory data concerning superstitious behavior. If an animal is given reinforcement on a totally noncontingent basis, it will tend to emit those behaviors which have by chance just occurred prior to reinforcement. For example, if we provide food to a hungry animal once every minute with no behavioral requirement and the animal has been circling impatiently prior to food delivery, there is a probability that circling will be increased. It is easy to appreciate this anthropomorphically by thinking about the number of superstitious behaviors that we repeat because some pleasant event has previously occurred in close proximity to this behavior.

I hope that this lengthy introduction will provide the reader with some appreciation of the reasons that it "seems so natural" for bears to beg. Are polar bears *(Thalarctos maritimus)* the sort of social animals that one might expect to develop similar behaviors under natural circumstances?

On the contrary, these largest living carnivorous land mammals live a solitary nomadic existence in the fragile Arctic ecosystem. They are strong swimmers, with special adaptations to their harsh environment. Although they have been popular in zoos and have suffered extensive hunting pressure, their behavioral patterns in the wild are largely unknown. It is established that in the wild, polar bears seldom seek conspecifics once fully grown (Larsen, 1971). According to Stirling (1974), unrelated bears would normally come into contact under only two conditions: food competition and downwind travel toward a sleeping conspecific. Males and females are in contact for a few days in April, the mating season. Seals constitute the major food source for the polar bear, although fish do represent a small fraction of the wild diet. Since fish are a nutritious and preferred portion of the captive diet, they were settled on as a primary food for our work.

GOALS FOR THE PROJECT

There were several ideas that we wanted to try simultaneously. First, we wanted to provide the bears a way to earn fish whenever they chose without having to beg from the public or wait for the keepers. Second, we hoped to demonstrate that, given time, most visitors would gain even greater enjoyment from watching the bears control things than from controlling the bears. Third, we wanted to increase the exercise for Esco-mo, the 10-year-old male, who had never maintained an appropriate physiologic reserve of fat. Our fourth goal—to reduce the consumption of rancid and inappropriate foods given by zoo visitors— had high priority. And, finally, we hoped to increase the varieties of behavior for Esco-mo and his mate, Iceter, who all too often were seen scrapping over choice morsels thrown by the public.

There has been little concentrated study of vocalizations in bears, with the exception of those which typically accompany aggression. However, vocalizations in the polar bears at Portland were found to be divisible into two major categories: hissing, which almost always occurred in the context of threatening or aggression, and a variety of other sounds seldom heard by the public such as "chuffing" (a low-intensity repetitive call) and a deep grumbling or roar best described as resembling the sound of a low chord on an organ. These latter vocalizations typically occurred when zoo staff entered the areas during nonvisitor hours. In contrast to Wemmer et al.'s (1976) suggestion that the chuffing sound is limited to stressful circumstances, we most frequently heard it in anticipation of feeding. In context, the deep bass noise seemed to be an apparent indication of recognition of zoo-keepers.

We reasoned that reducing sounds attendant to aggression might help with our task of reducing conflict in the grotto. At the same time, we thought it would be nice for the public to hear that the bear was capable of other sounds. So a special transducer was designed which was responsive to all vocalizations except hisses.

PROCEDURE I: THE BEST-LAID PLANS . . .

There was a long, cool corridor extending behind the polar bears' grotto (Figure 6-1). For ease of loading fish, staying out of the rain, and minimizing refrigeration needs, we installed our apparatus in this corridor. The feeder consisted of a conveyor belt and catapult system which were linked through instrumentation to the voice-operated microphone system (Figure 6-2).

The bears were shaped by successive approximation to approach the slit window in which the microphone was installed. After successfully accomplishing this step, vocalizations of any sort other than hissing were reinforced when either bear vocalized near the window. Finally, the procedure was fully automated with the gain on the microphone adjusted to require clearly audible sounds. A soft, distinctive tone which could be heard throughout the grotto served to alert the

Figure 6-1. Long, cool corridor housed original feeder belt and catapult device for bears. *(Photo by R. Fial)*

Figure 6-2. Voice-operated relay system used to detect and filter polar bear vocalizations. *(Photo by R. Fial)*

bears when food was available should they wish to order it. The interval between the stimulus and the previous fish delivery was 15 seconds, and the brief tone was repeated at 30-second intervals until vocalization occurred.

Successful vocalizations resulted in the sounding of a bell and catapulting of fish a considerable distance into the exhibit. The opening for fish delivery was approximately seven meters from the microphone. At first, things looked good. Activity was effectively increased, begging was greatly reduced, and both the residents and the zoo-goers seemed to appreciate the new opportunities. But then Iceter developed a new behavior which illustrated the incompleteness of our planning and the inadequacy of our design.

She would approach the food delivery area and stand with her mouth open while Esco-mo ordered the food. Obviously, this did not reduce levels of aggression (Figure 6-3). We shut down the apparatus and went back to the drawing board.

Figure 6-3. Polar bears *(Thalarctos maritimus)* lock jaws in dispute over fish. *(Photo by H. Markowitz)*

PROCEDURE II: TWO BEARS, TWO FOOD SOURCES

The following year, we completed new equipment which would deliver food from high on the roof behind the grotto (Figure 6-4). This apparatus included two systems operated simultaneously. One segment delivered fish near the area where "the order had been placed" and the other tossed fish into the pool in the center front of the grot-

Figure 6-4. Revised polar bear food-delivery system mounted on grotto roof uses horizontal catapult to throw fish into pool. *(Photo by N. Lee)*

to. The microphone, associated stimuli, and intervals were the same as for the previous procedure.

This food-delivery modification satisfied the goals originally proposed for the experimentation, and it simultaneously provided an opportunity to observe interesting changes in the polar bears' behavior and in that of the zoo visitors. In the earlier experiment, the majority of responses (89.7 percent) were accomplished by Esco-mo. In many cases, he would simply push Iceter out of the way, order fish, and then lumber over to pick them up. Although he ordered nearly 90% of the time and consumed the majority of food, Esco-mo pursued and hissed at his mate when she finally adopted a method to get her fair share by standing near the catapult and catching fish on the fly.

Altogether, Esco-mo had ordered more than 4400 fish using the first paradigm, and we expected to see him continue responding. We thought that he might consume the fish delivered near the microphone and leave the catapulted fish for Iceter. Instead, beginning with the first day of the new installation, there was a surprising change in the male's behavior. He would sometimes move quickly away from the "order window" and dive into the pool to get the fish delivered there. Within the first two days, Iceter earned 20 percent of the food. By day three, she ordered 75 percent of the food, and throughout the remaining 68 days of the procedure, in excess of 95 percent of the more than 5000 orders were placed by Iceter.

In the meantime Esco-mo was apparently having great fun. He would move up to the precipice above the pool and high-dive in (Figure 6-5), submerge, and then come up with food. It was especially nice to see him evolve these behaviors on his own. Indeed, the one criticism from a visitor about this work was from a person who thought that some evil researchers had compelled the poor bear to swim. This may indicate something about what begging bears in zoos do to educate the public about the species . . . a largely aquatic mammal is now seen to be doing the natural when it begs and the unnatural when it swims.

There were other interesting anecdotal outcomes, one of which is especially memorable. A pleasant young woman came running up to the bear grotto with a batch of her friends trailing behind. She began to entreat Esco-mo to pay attention to her by saying, "Esky! Esky! . . . Come!" and waving a marshmallow at him. But Esco-mo kept diving in the water for fish. I have to confess that it was probably bad judgment, but I could not resist leaning over to softly tell her how much fun it was to see the public begging the bears for a change. Fortunately, she took this with good spirits, blushed a little, and volunteered to help with the work.

Figure 6-5. Polar bear dives into pool to consume fish. *(Photo by J. Mellen)*

CONCLUSION

There are dozens of changes that we would make in designing a permanent installation, especially with the opportunity to work in the construction phases of a new polar bear exhibit. These animals are at their most beautiful when they are swimming, and exhibits such as that at the Calgary Zoo, which maximize opportunities for both underwater and surface viewing, are much more educationally ideal. It might also be desirable to provide the opportunity for a more naturalistic hunting behavior rather than vocalization as a response requirement.

Yet, this experiment illustrated that a number of ambitious goals were readily accomplishable within the available environment and budget: activity was increased, visitors gained appreciation for a wider variety of behaviors by the bears, including a greater proportion of species-typical behaviors such as diving and swimming, consumption of junk food was sharply reduced, and aggressive vocalizations and aggression were diminished. The zoo's veterinary staff began to report significant changes in the male bear's physique accompanying the active exercise (Schmidt and Markowitz, 1977; Markowitz, Schmidt, and Moody, 1978). Esco-mo put on the normal reserve of fat, and there were few aggressive bouts through the remainder of the work until Iceter was denned and successfully mothered a youngster named Cheechako.

As with most of our work, the research staff's major rewards were watching the bears' responsiveness to the opportunity for autonomous feeding and the visitors' increased appreciation of the animals' capabilities. People still crowded around the exhibit and occasionally tossed things to the bears. But, for the most part, they were interested in knowing about why the bears were vocalizing, diving, swimming, and catching "flying" fish. It beats the hell out of begging for a living.

7
DISCRIMINATION STUDIES

A prominent theme in the comparative animal literature has been the attempt to assess relative capabilities in discrimination learning. Sometimes this has been accomplished with the assumption that comparative measures of intelligence are being made. However, most researchers in this area have tended to increasingly disallow this interpretation. For example, Duane Rumbaugh, in his extensive studies of the relative abilities of primates, wound up with results that placed the gibbon near the bottom of the range for those animals investigated. But as Rumbaugh suggested, the probability is that the task utilized was much more appropriate in terms of required responses for the other animals than for the gibbon. In a similar fashion, some early studies of capability in rodents—supposedly descendants of bright animals—illustrated that this was a pure function of the type of task involved. We described this in some detail in an article called, "The superiority of maze-dull animals on visual tasks in an automated maze" (Markowitz and Becker, 1969).

Despite the fact that it is clear that the term "intelligence" may be overused to the point of being valueless, there do remain a lot of interesting questions about the way in which different species attack particular problems. In this chapter, a variety of studies involving light-dark discrimination with harbor seals, gibbons, orangutans, and elephants will be described. Although these will do little to allow judgments of relative capabilities because of the differences in milieu, they should help the reader to gain some feel for the variety of methods by which different animals may solve problems of a similar nature.

HARBOR SEALS

Early in the work at Portland, we began to study harbor seal *(Phoca vitulina)* discrimination learning in a somewhat traditional fashion. A testing chamber was built (Figure 7-1), and three press panels and a slot through which fish were automatically delivered were installed. The chamber was transparent on the side facing the water so that seals who were not working could observe the one in the chamber.

As anyone who has worked with pinnipeds will attest, seals and sea lions are extremely capable and cooperative subjects. It has often occurred to me that the only reason that they are not described with the glowing terms used for cetaceans such as the dolphin is because pinnipeds have not had as vocal champions for their cause. We have

Figure 7-1. Harbor seal *(Phoca vitulina)* in testing chamber. *(Photo by B. McCabe)*

been told so frequently of the massive brain and potential capabilities of the dolphin that it has often been overlooked that the seal has virtually the same proportion of neocortex as do the dolphin and man. In the study reported here, we found the harbor seal to be an extremely willing and independent worker, capable of setting up all sorts of "personal" ways in which to solve problems.

Each day, experimenters would go to the island in the center of the seal pool, where the chamber was installed, to run several hours of research. The seals were coaxed to haul out of the water individually and to enter the chambers. While some of them were a little bit fussy about being closed in, most of the seals adapted very well to the situation after a few days. We began to collect traditional light-dark discrimination data by randomizing the side that was lighted and rewarding the animal for responding to the appropriate stimulus. While they were learning to do this, I found myself becoming increasingly disconsolate with the testing paradigm and, for largely selfish reasons, redesigned the seals' activities.

By this time, it had become clear to me (Markowitz, 1973, 1974) that there were great advantages from both the animal's and researcher's standpoints in allowing as much freedom as possible in the solution of problems. My happiest times with the seals were spent watching my friend and student, Brian Johnson, wade in with them, or watching the seals swim about and play with each other before and after the testing sessions. The chamber was dismantled and in its place we installed two large globes separated by about eight meters. These globes, which hung a few inches above the water's surface, incorporated both visual and auditory stimuli and could be moved by the seals to indicate a discrimination choice (Figure 7-2). The new research protocol involved allowing the three harbor seals simultaneous access to this apparatus. We first coaxed them with fish to demonstrate that tilting the globe would lead to reward. Each seal learned this in a matter of minutes. Thus began a series of studies which continued for six years.

A light-dark reversal discrimination paradigm was adopted. The essentials of the procedure were that one of the globes would light at random and would simultaneously make a distinctive noise. (The noises were added to ensure that the stimuli were easily discernible on bright days, since even aircraft lamps in the globes were not sufficient to guarantee contrast with intense sunlight.)

Figure 7-2. Apparatus allows harbor seal testing in a free-swimming group situation. *(Photo by H. Markowitz)*

The seals learned first to always go to whichever globe was illuminated. After they had accomplished this to a criterion of 20 consecutive correct responses, the rules were reversed. Now the seals had to always go to the dark, silent side rather than the noisy, lighted globe.

If the results of this work are viewed strictly from the standpoint of performance on the task, there is nothing terribly surprising. The seals showed the typical great difficulty on their first reversal, which is seen in most reversal discrimination studies (e.g., Essock and Rumbaugh, 1978; Markowitz and Sorrells, 1969). They did not approach consistent one-trial learning of the sort that Harlow and others have described for individual organism designs. This was in keeping with what we have found for other complex species in the noisy zoo milieu, where there is access to the apparatus by more than one animal (e.g.: Davis and Markowitz, 1978). Instead, the great richness of the results lies in the observation of the interactions between the seals in accomplishing their daily work.

In general, this changed regimen was a success with experimenters and the public from the outset. Visitors could come to the zoo any day during the regular sessions and be guaranteed that they would see active, healthy seals whose diet was being carefully monitored. After initial training in which each of the seals learned to respond to the lighted globe, the apparatus was simply turned on daily and experimenters recorded the seals' behavior and delivered fish when appropriate responses were made. In the beginning, these fish were tossed approximately 20 feet from the correct response and on the side where the correct response was made. Here is the way the seals handled the problem.

Although Milhouse, who was two years old, had learned to perform light-dark discriminations in the chamber, and we expected that we might see some transfer to the new task, Neptune, one of the two untrained year-old seals, began to work the first session and others soon progressed to poach on his earnings. Despite the fact that Olin was trained by successive approximation to hit the globes, he never worked during the experimental sessions, but learned quickly where the reinforcers were delivered. Olin illustrated his expertise in discrimination by swimming to the correct side and waiting for the fish to drop, while Milhouse and Neptune raced from the correct manipulandum to contest for the reward. Following the solution of the first free-swimming discrimination problem when the condition was changed to dark correct, Neptune made a few errors and then resigned from the task, leaving the work to Milhouse. For a brief time, the results suggested that we might be observing "specialization" of the sort that will be described below for discrimination studies with the gibbon. However, part way through the third reversal, Neptune stopped responding entirely and adopted Olin's strategy of letting Milhouse do the work and "helping" to eat the fish he had earned.

It was at this point that we began to see some beautiful illustrations of the seals' ingenuity. Milhouse developed a marvelous decoy behavior with which he was able to lure the other seals to swim with him as he nonchalantly ignored the correct stimulus. This behavior involved repeatedly swimming alongside them until they finally began to play and accompany him. When they got to a distant part of the pool, Milhouse would suddenly porpoise out of the water, race in the opposite direction, and hit the globe at full tilt on his way to catch the fish. Eventually, Neptune and Olin refused to follow Milhouse's diversionary tactics and began to gather much more than their share of the smelt.

Consequently, we changed the feeding protocol, and used a table of 100 successive random presentations in which food was unpredictably thrown to one of three areas in the pool. This revised procedure was used through all of the remaining work to be discussed here, and it did result in an even distribution of fish regardless of which seal was doing the majority of work. The chance to see behavioral individuality was hardly at an end, however. During the ninth reversal condition, an unexpected visitor entered the scene. Each day as the feeding procedures began at regular times, a herring gull *(Larus argentatus)* alighted on the roof of the pool access room. His distinctively shaped beak and unique behaviors made this gull, whom researchers named Jonathan, easy to identify. From his perch above the seal's work, Jonathan would rock slowly back and forth while watching them solve the discrimination problem. Without warning, after several sessions of observation, he swooped down at precisely the right moment to filch a fish from the section of the pool which Milhouse had come to treat as his feeding domain. Milhouse watched the fish disappear, then raised the front part of his torso from the water and rocked aggressively, finally hauling out onto the island. The apparatus automatically switched to the next condition, and when Milhouse did not respond, Neptune swam over, put his head and neck near Milhouse on the island and repeatedly poked at him, seemingly urging a return to work. But Milhouse would not reenter the water and, after several fruitless attempts to convince him, Neptune took over the task. It was clear that his failure to respond for the previous five weeks had not resulted from forgetting how to work on this paradigm. Neptune made 21 responses in a row with little hesitation before Milhouse rejoined the activity.

Throughout much of the period represented by Figures 7-3 and 7-4, Milhouse did most of the work, but the graphs are based upon group data. The rapid decline in trials to criterion for the first 80 reversals mirrors results from many of the more traditional isolation studies of discrimination reversal ability (e.g., Gossette, 1973; Markowitz and Becker, 1969). The abrupt increment in trials to criterion, beginning with the sixth group of reversals, is related to the shifting of location from one pool to another. The apparatus was identical and spacing was approximately the same, but the second pool was larger in diameter and much more shallow. Meanwhile, the original pool was modified and entirely enclosed in glass, wood, and concrete, significantly diminishing the distractions which visitors imposed on the seals. The seals were returned to this pool during the seventh group of 20 reversals, and they continued progressing in efficiency.

However, the zoo changed directors, and an administrative decision dispossessed the seals from their new home and returned them to the pool where thoughtless visitors occasionally threw in dangerous objects that were consumed by these trusting animals. At least we were able to have the pool altered by increasing the depths so that it now contained about 370,000 gallons of water. As shown in Figure 7-3, there was an immediate jump in trials to criterion and another gradual improvement through the 15th series of reversals.

Finally, beginning with the 16th reversal group, a novel behavioral situation led to an additional retracing of the acquisition curve. An inexperienced seal, who had been introduced to the group, began to respond for the first time. As Tillamook took over more of the work load, the payoff efficiency was decreased. The final reduction to levels of 150 or fewer trials to criterion represents some learning by Tillamook plus an increased return to predominant responding by Milhouse. For those readers interested or familiar with the traditional literature on discrimination biases, Figure 7-4 has been included. This figure indicates that the seals were showing no significant side preference in their performance. The curve for left correct responses is virtually identical with that for total correct.

Emerging from a complete analysis of these data was the fact that every retracing of the acquisition curve was related to identifiable

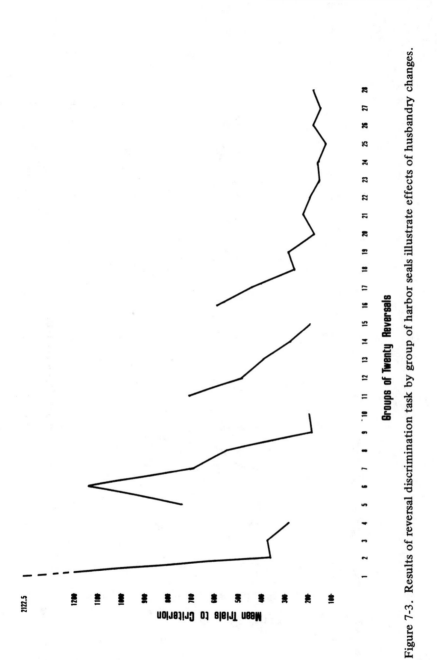

Figure 7-3. Results of reversal discrimination task by group of harbor seals illustrate effects of husbandry changes.

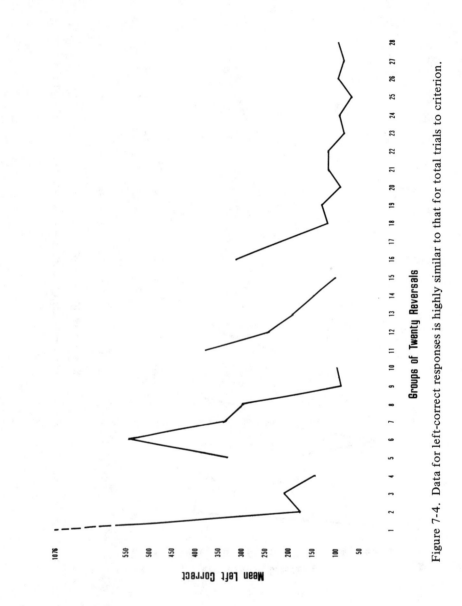

Figure 7-4.   Data for left-correct responses is highly similar to that for total trials to criterion.

husbandry changes in the zoo. This illustrates that the harbor seals' performance was not simply a function of learning stimulus contingencies, but also represented a general adaptation to environmental changes. It also serves to underline that data from learning tests run in noisy environments, although difficult to analyze, may produce systematic results that have more bearing on common life situations than studies run in isolation (Markowitz, 1978). Had we accomplished this work by continuing to use the isolation chamber, our most interesting and informative results would never have accrued.

GIBBONS

Two white-handed gibbons *(Hylobates lar)* mentioned in Chapter 3, Venus and Milo, had a simple discrimination panel placed in the window of their home cage (Figure 7-5). The transluminated press panels were programmed to provide the same standard light-dark reversal procedure described above, and food, including monkey chow and fruits, was automatically delivered for correct responses. Since there was a relatively restricted area in which responses could successfully be accomplished, the only really unusual features of this paradigm compared with laboratory procedures were: (1) the gibbons could choose which of them would respond, (2) there was no real deprivation and, as with our other projects, feeding was simply forestalled to the regular time if they chose not to respond, (3) the apparatus was available for long periods of the day, and (4) the noisy zoo environment occasionally included people rapping on the windows and sounds from adjacent cages.

With these two gibbons working together, we found that the male learned the light correct problem very quickly and the female took considerably longer (Markowitz and Woodworth, 1978). But when the conditions were reversed, Venus eventually accomplished the task, while Milo showed no progress after a very long period. It was arbitrarily decided to run a "cage criterion" in which any 20 correct responses in a row, regardless of which animal was working, represented success. As shown in Figure 7-6, the gibbons continued to become somewhat specialized as work progressed through 18 months for more than 1300 reversals.

Figure 7-5. White-handed gibbon *(Hylobates lar)* responds to simple discrimination task in window of home cage. *(Photo by H. Markowitz)*

Several times in the progress of this work, I had to dissuade researchers who were anxious to "shape up" Milo's performance. It always seemed to me that we would have lost much that was valuable by this excessive education. Once animals have learned the basic skills necessary to address problems, it is both more challenging and

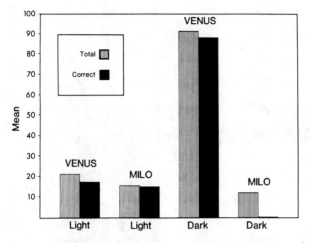

Figure 7-6. Mean number of correct responses and total responses for two gibbons with free access to discrimination panel.

entertaining for them and more instructive for us to observe progress rather than forcing it on them.

ORANGUTANS

The apparatus that was used in the study with gibbons was later moved to the enclosure where a breeding pair of orangutans *(Pongo pygmaeus)* named Harry and Inji resided. Although both of these animals were also given equal access to the equipment, the male made more than 99 percent of the responses (Davis and Markowitz, 1978), so most of our discussion will center on his behavior. One prime reason for conducting this work was to provide activity for the orangs while we prepared the more complex entertainment of tic-tac-toe for them (described in the next chapter). If we had planned a long-term reversal discrimination study, the apparatus would have been changed to facilitate Inji's responding since there was every evidence that she wanted to participate. Her problem was that the response panel was near the top of the door and the reinforcer was delivered toward the bottom (Figure 7-7). When she did choose to work, it required some speed to grab the reinforcer before Harry snatched it. In contrast, the male with his much longer arms could sit on the floor, raise him-

Figure 7-7. Orangutan *(Pongo pygmaeus)* responds to light-dark discrimination task. *(Photo by J. Mellen)*

self up slightly to respond, and sit back down in time to meet the reinforcer being dispensed.

Figure 7-8 illustrates the average trials to criterion for the orangs in this task. In light of the results described above, there is nothing terribly surprising here. Once again, systematic results are obtainable without the isolation of the laboratory, and the goals of increasing activity and providing diversion for the orangs were accomplished. Harry was willing to respond through more than 1000 reversals, indicating that even the simplest artificial apparatus can provide stimulation when compared with traditional zoo routines.

These results also indicate to us that the level at which asymptote is reached is not merely a function of the animal's complex capabili-

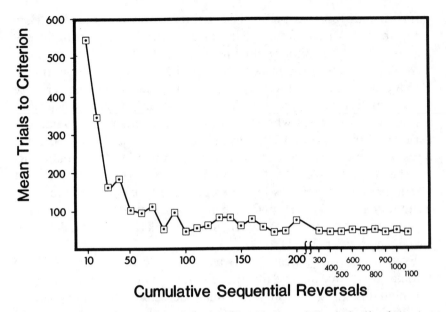

Cumulative Sequential Reversals

Figure 7-8. Curve of average trials to criterion in reversal discrimination by orangutans. *(Diagram by Diane Fenster)*

ties; it is also clearly related to the complexity of the milieu. Surely, orangs are as capable as any species with which we have worked and, in a sound-proof chamber that eliminates all outside distractions, there is little question that they might quickly reach the pinnacle of "one-trial reversals" described in the traditional literature. But, these conditions are never ordinarily approximated in an animal's everyday life. These results, which showed leveling at approximately 49 responses per reversal, may best be evaluated by looking at the average relative noise in the environment where discriminations were being accomplished. This is exactly the way in which I would propose we should always progress in making evaluations of an animal's success in coping with its environment. Except for highly artificial laboratories or libraries in which *Homo sapiens* may wear earphones and stick their heads in cubicles to read, life without distractions seldom exists.

ELEPHANTS

The major results of our light-dark discrimination test in the Asian elephant *(Elephas maximus)* have to do with long-term memory and animal health topics which will be deferred for a later chapter. Here I will touch on two factors of importance for formal testing of elephants: their tremendous strength and lack of great patience.

While elephant cows in particular can be extremely gentle, tractable, and even loyal to their regular human companions, the very size of the animals may represent a potential hazard. Constant vigilance is required of workers who come into direct contact with them. Experimental protocol that requires disproportionate attention to apparatus adjustment, refocusing, etc., is quite impractical where elephants are the subjects. We have found that a very convenient way to test discrimination capabilities in the elephant is to use the method pioneered by Leslie Squier: place the apparatus outside the animal's enclosure so that it can only reach it with its trunk (Markowitz, Schmidt, Nadal, and Squier, 1975).

Even where this precaution is used, great care is required if the testing procedures involve fine discriminations or excessively long experimental periods. On one unfortunate occasion when our apparatus was accidentally placed a few inches closer than usual, an elephant impatient with lack of early payoff for its work began to topple the discrimination box, which weighed more than 100 pounds and had 200 pounds of investigator inside. The 300 pounds was moved with an effortless flick of the elephant's trunk. Another time, a joint between one-inch sections of plywood that had been glued and screwed together was pulled apart as if it were stuck with library paste. Even where steel, slate, and other tough materials are utilized, the best precautions include careful attention to detail and procedures that minimize the probability that the elephant will become impatient. To illustrate this point, I will describe in brief detail a system that we designed to screen for visual acuity in the Portland elephants.

We wanted to measure minimum separable acuity and knew that standard psychophysical laboratory procedures requiring thousands of trials of repeated testing would simply not work. So, we arranged a simple protocol in which a few hundred trials a day were sufficient to train the elephant for the critical testing day. A ground glass screen was placed in the middle of our testing apparatus, and beneath it two heavy-gauge steel disks were installed (Figure 7-9). We produced a series of vertically ruled slides with spacing varying from only two lines for the entire slide to gradations much finer than those separable in human acuity. Neutral density filters were used to equate the

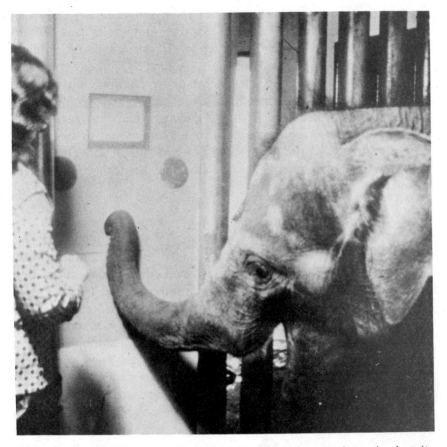

Figure 7-9. Young Asian elephant *(Elephas maximus)* responds to visual acuity test. *(Photo by H. Markowitz)*

amount of luminosity for each slide. The pretraining, the last step accomplished prior to the writing of this chapter, went smoothly by using just two slides. One of these was a very broad grid and the other was a clear neutral density filter that transmitted the same amount of light. Temba, the young elephant used in this training, learned to always go to the left for a grid and to the right for a solid. In a matter of a few weeks of testing, this response was accomplished with better than 95% reliability.

The next anticipated step in this type of testing is to randomize 100 presentations, 50 of which involve the old training stimuli, while the other 50 represent all of the other separations in our total range.

The use of 25 solids and 25 original broad gradations is to allow an assessment of continued stimulus control. If the elephant fails to perform at close to the 95% accuracy level, we know that it is failing to respond in the expected manner. As long as stimulus control is maintained, our first approximation of discrimination ability will be that those lined stimuli treated as solids are beyond the minimum separable acuity threshold of the elephant. While this leaves some technical questions about whether true maximum sensitivities are being assessed, we propose the method as a good initial screening device for animals such as the elephant where more lengthy procedures are impractical.

CONCLUSION

A little library research will quickly convince the potential zoo researcher that not much is clearly established about the discrimination capabilities of most exotic species. Careful attention to the special requirements for each animal will help to ensure a welcome from zoo administrations that might otherwise be wary of such procedures. Where permission can be obtained to make measurements on a chronic, in-milieu basis, especially intriguing and unexpected outcomes may be forthcoming. With some of the earlier examples described in this chapter, I have attempted to illustrate that these procedures can simultaneously provide entertainment and novel stimulation for the captive animals. In this sense, discrimination studies can clearly fall within the scope of behavioral enrichment in the zoo.

8

GAMES ANIMALS PLAY

The mandrill *(Papio sphinx)*, largest of the monkeys, is a mostly terrestrial primate that deserves to be housed in a spacious exhibit. Unfortunately, the Portland mandrills lived in a stark concrete cage. Bright blue and red coloration on face and rump make mandrills easy to identify and popular in zoos. There is significant sexual dimorphism in the species, with the male a great deal larger and stronger than the female. While little is known about the wild behavior of the mandrill (Rowell, 1972; Jouventin, 1973), it seems apparent that much of the captive behavior may be directly related to space limitations. The exhibit enclosure in Portland was 8.8 meters wide, 6.5 meters high, and 5.5 meters deep.

Most striking in the behaviors of our male mandrill (Blue) was the fact that he would occasionally leap from the back wall to the front cyclone fence when keepers or observers whom he knew came by. Blue would shake the fence vigorously and, on more than one occasion, actually tore part of it out. His behavior with respect to the female mandrills was usually exhibited in brief, aggressive spurts. Blue also cowed them with occasional shrug threats and head-shaking, and, in general, greatly limited their cage usage (Yanofsky and Markowitz, 1978). We wanted to find some stopgap solution that would allow the females more complete use of the space, entertain Blue, and perhaps give the public some idea of the mandrill's quickness. Bill Myers (1978) had reported some work performed in the back room of a zoo using a baboon as a subject in a speed game. We decided to modify this approach to allow Blue access to the game throughout the day whenever he wished and to guarantee him a game whether there was an experimenter or zoo visitor around or not.

Two identical consoles were installed at the mandrill exhibit. The one for the mandrill was located in one of the doors to his cage (Figures 8-1a and b) and the one for the human was installed so that pass-

Figure 8-1a. Mandrill *(Papio sphinx)* sits at console ready to play game against visitors. *(Photo by G. Stryker)*

Figure 8-1b. Mandrill playing speed game. *(Photo by H. Markowitz)*

ersby could watch the progress of the game and the relative performance of the competitors (Figures 8-2a and b).

The game was very simple with rules as follows: first, the large disk in the upper left-hand corner lit up for the mandrill. This signal remained on indefinitely until the mandrill chose to play, since the game had been designed primarily for his entertainment. Blue's response to the lighted disk turned on a similar disk on the public's panel that invited visitors to drop in a dime if they wished to compete. However, if no one deposited a dime, the mandrill was always able to play because the computer would then compete with him. Once the game was initiated, a square would come on in one of three random locations, simultaneously on both panels. Whoever touched the light first won. Premature responses were recorded as automatic losses. The first competitor to gain three wins was rewarded, the mandrill with a piece of food and the visitor with an enunciation on the scoreboard showing that he or she was quicker than a monkey.

Throughout the above discussion, I have talked about the game as if it were Blue's exclusively. The truth is that in theory it was avail-

Figure 8-2a. Human *(Homo sapiens)* contestant competes with mandrill. *(Photo by B. McCabe)*

able to all of the monkeys in the cage, but the male virtually never let anyone else approach "his" game. His progress in training astounded us. We began by simply turning on one of the three squares and reinforcing the mandrill for touching it. Then we progressed to randomizing the position of the light. Blue made the transition between these responses without error. We next began to require that he make his response within a minute and finally, began to turn on the disk so that Blue could initiate the games when he wished. Probably because all of the responses he had been making were to lighted stimuli, Blue picked up his new requirement without hesitation in a single trial. In a matter of weeks, he was ready to compete with visitors.

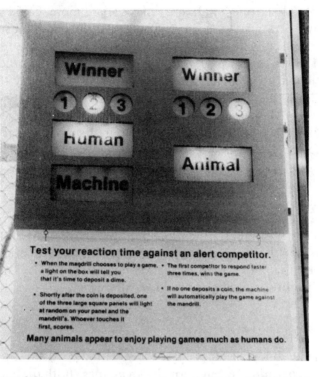

Figure 8-2b. Scoreboard for speed game. *(Photo by B. McCabe)*

We knew from previously reported studies, such as Myers's, that the monkey might be expected to compete well with people in this situation—but we never anticipated speed of reaction as quick as Blue's. He stabilized at winning more than 70 percent of the games in fair competition with visitors. This would have surprised us more if we had not been simultaneously monitoring his performance in competition with the computer. The computer essentially worked on a titration schedule. That is, when Blue won more than 70 percent of the games against the computer's reaction time, the computer decreased its time by .005 seconds. Blue continually progressed until he was able to compete with times faster than 310 milliseconds. Typical reaction time to a single stimulus in a known position for humans, chimpanzees, and rhesus monkeys (the most often studied primates in this situation), is about one-quarter of a second. Here was a primate who, in six one-hundredths of a second more, could scan three alternatives and make a correct selection. No wonder most of us did not often beat him at his game.

All of us would have preferred a big open corral or other naturalistic exhibit for the mandrills, and we hope that someday this will be provided. In the meantime, the game had accomplished much of what we had hoped for. The mandrill apparently enjoyed earning his food this way and almost never required supplementary feeding. Most afternoons we had to hang out a sign saying, "We don't want him to get fat." The sign explained that the veterinary staff controlled the maximum amount that the animals were to eat. The females' cage usage was dramatically increased, and they were in general much less "spooky." The public loved the game, and it was featured in a number of national media spots that gently teased reporters about their inability to consistently beat the monkeys. We had hoped that this would be a good temporary display. Blue transformed it into something at which we all marveled.

ORANGUTAN TIC-TAC-TOE

The general consensus of primate lovers in the Portland Zoo was that the worst conditions existed for the great apes. Both the chimpanzees and the orangutans were in fairly small, dark cages of the same concrete construction described above. Instead of having cyclone fencing, the front of their cages was made of extremely strong glass. As a diversion for the orangutans, we decided to install a tic-tac-toe game. We built panels for both the public and orangs with the intention of installing them in a fashion similar to that described for the mandrill speed game.

Because of the political changes that caused us to end the program in Portland, progress was only completed through the first stage of this work, and the public never had a chance to compete with the orangs. The training in itself did provide some interesting data, and considerable time was spent in this systematic program that would give the animal the most flexibility in learning (Figures 8-3 and 8-4). In the first stage of training, we simply shaped the orangs to respond to the "I want to play" light and then to the "O"s on the game board. The first program followed the rules of tic-tac-toe, but did

Figure 8-3. Orangutan *(Pongo pygmaeus)* next to tic-tac-toe game. *(Photo by G. Stryker)*

not win until no other alternative was available. A little reflection on the restricted number of moves possible in tic-tac-toe will make it clear that losing is not always as easy as it sounds. As a matter of fact, a "try to lose" game often defeats young children.

After Harry, the male orangutan, had progressed to the stage where he beat this intentionally inept "opponent," the next step involved a random program that followed the rules. This program neither "intentionally" tried to win nor to lose. There was every reason to believe, based upon Harry's rapid progress and the general abilities of this complex species, that eventually the orang would seldom have lost against human competitors.

Figure 8-4. Orangutan closely scrutinizes response alternatives in tic-tac-toe. *(Photo by G. Stryker)*

A SPIDER MONKEY SPEED GAME

In Chapter 10, some major naturalistic apparatus produced for the Panaewa Rain Forest Zoo in Hilo, Hawaii, will be described. Although our major emphasis in the Hilo work was to develop methods for tigers and gibbons to hunt and gather food in an acre-size exhibit, we also did a smaller project for the Children's Zoo. This area was essentially a holding station for a small number of animals prior to their introduction to the larger exhibits, but it did have some interesting features. The building was constructed so that the animals living there could select to be undercover or out in the weather (Figure 8-5). This was easy because the temperature in Hawaii never required any artificial heat. During one of his trips to Portland, Jim Juvik, who con-

Figure 8-5. Children's Zoo building in Panaewa allows animals to choose indoor or outdoor exposures. *(Photo by J. Juvik)*

tracted with us to work in Hilo, had a chance to play the speed game against Blue, and he asked us to produce an almost identical game for mandrills in the Hilo children's zoo. The only difference was to be the fact that the public could interact by pressing an "I want to play" button similar to the animal's, rather than dropping in a coin as they had done in Portland.

While several of us were diligently producing the parts in preparation for installation by our summer class, the building plans underwent some changes. Final construction details clearly made the cage much too light in construction to adequately house mandrills, and at the last minute we were confronted with a new challenge: a pair of spider monkeys *(Ateles ater),* most unlikely candidates to successfully compete in such a contest. Where the mandrill had been a natural for using its digits to push at things, the spider monkey was much more arboreal with hands specialized for swinging. With the promise that we could agree upon another animal if things failed, the class decided to see what would happen if we placed a rope near the apparatus to attract the monkeys to the area and encourage button-pushing (Figure 8-6). The male did little responding, but the female quickly became adept and in her own unique way. Instead of leaning over and pushing the buttons, she would hold onto the rope and press them

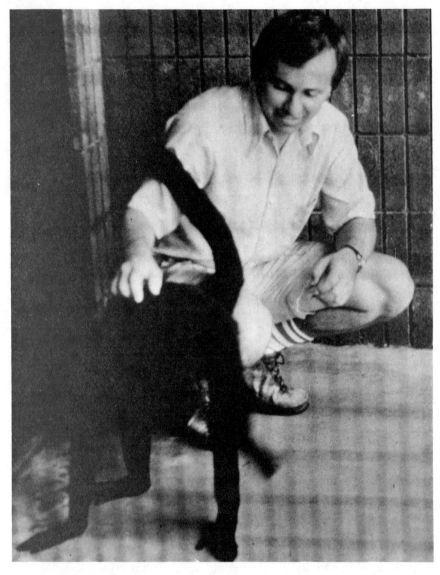

Figure 8-6. Jim Juvik and spider monkey *(Ateles ater)* inspect speed game. *(Photo by R. Fial)*

with her nose. Despite this awkward-sounding posturing, her speed was sufficient to beat many visitors.

Because faculty members and students from the University of Hawaii had expressed an interest in continuing the work when we left, I taught a summer course at the Hilo campus, and we made special provision for continued data collection. This included both electromechanical counters and an output which could be directly plugged into on-line recorders for transport to the computer (Figure 8-7). Once we found that the game would not be used exclusively for mandrills, stress was placed upon adjustable response requirements. Different animals temporarily housed in this facility will be able to use it to pass the time and teach researchers something about their abilities.

SOME FINAL THOUGHTS ABOUT GAMES

I have already discussed the fact that games do not seem to me to be ideal zoo exhibits in spite of their great popularity with the public and the media. Increasingly, I find myself advising my students and

Figure 8-7. Panaewa Zoo speed game controls and data-collection devices.

colleagues to spend more time, raise more money, and try to make more naturalistic opportunities for the animals. But, given exactly the same circumstances and impoverished environment, I might very well provide the same amusement for the mandrill or orangutan today. There are some difficult philosophical questions about when such temporary procedures should be used.

In developing personal guidelines to try and deal with these questions in my own thinking, I have found it necessary to use a mixture of the empirical and the anthropomorphic. Empirically, it seems to me that if the animals that are provided the game are not deprived, but receive their food at the same time as their neighbors even if they choose not to play, then the game is a luxury rather than a compulsive requirement. Similarly, in each of the games which we installed, the animal was intentionally given ad-lib opportunity to initiate the game at its whim. There was no time limit between the first lighting of the "I want to play" button in the morning and the animal's initial decision to participate. Although it sometimes turned out that one animal became possessive with the game (such as with Blue), we intentionally installed the apparatus so that all cage members might have a chance to use it.

Taking an additional data-based look from the standpoint of results of the mandrill game, Blue seemed clearly to prefer to compete with the public as opposed to games against the computer. This was true for slow computer speeds as well as for those that became faster than the average visitor. Our daily records show that when no one came to play against him, Blue would often leave the game and go about other activities.

Blue's gestures and other expressions—only anecdotally summarized in the data that we have collected—lead us more into the area of anthropomorphic conjecture. He would turn and appear impatient when people did not immediately elect to play the game against him. After contests that Blue won, he would typically turn and look down his nose at the competitor after accepting his food reward. When he lost, there were frequent shrugs and even an occasional rapidly nodding threat. I am not embarrassed to say that I think he really "got off" on the game in the same sense that human competitors often do. The final point is a closely related one: the circumstances in which we designed and implemented games were ones in which any of us would have welcomed the opportunity for any variety in an over-whelmingly barren environment. Games are fun when you don't have anything more exciting to do.

Occasionally, we have had what I think are innocent misunderstandings from people who believe that we have built apparatus and trained animals to give the appearance that they are competing in human-like fashion. But this is clearly untrue. The animals with which we worked were shaped for only the briefest period so that they could learn how to play the game *if they wished*. Further, they were not acting *as if* they were competing; they *were* in fact playing on an equal and fair basis. Some days they chose to play more in the morning and some days more in the afternoon—no one ever insisted that they "show off" for visitors. From my perspective, this was what made things fun for all concerned, including the animals, and distinguished the work from carnival acts where highly deprived or rigidly controlled animals are forced to go through stereotypical routines.

9

ON ELEPHANTS FORGETTING
AND BATHING

The Portland Zoo is famous for its wonderful herd of Asian elephants *(Elephas maximus)*. Thonglaw, the first herd sire in the western hemisphere, has now passed on, but his offspring Packy continues the active breeding program in this facility. While it is not as large as the African elephant *(Loxodonta africana)* where the bulls represent the largest living land animals (Schmidt, 1978), the Asian elephant is also impressive. Bulls may reach a height of 3.2 meters at the shoulder and a weight of 5400 kilograms. Average Asian elephant cows are about 2.3 meters in height and weigh approximately 3000 kilograms. In spite of this awesome size, elephants remain remarkably sensitive and, in some ways, delicate creatures that all too frequently suffer from inadequate captive care.

Given the opportunity, some elephants will take care of much of their own foot and skin care, but this requires areas with abrasive materials that are not always available in typical zoos. For many of the elephants in the typical contemporary parks, routine pedicures are an essential treatment and many facilities are inadequately outfitted to safely allow this necessary care. One way in which behavioral researchers may be of considerable help to the zoo staff is in designing training protocol. At the request of several of the keepers in Portland, we worked out a schedule that allowed them to train a most difficult elephant to allow each of her legs to be independently chained and unchained without moving. This made it possible to more safely administer health care and enhanced routine husbandry procedures.

The first step in this program involved getting all the personnel who worked with the elephant to use the same jargon and the same intonations. Then, by simple successive approximation, we worked in short daily sessions to reach our final target of keeping all legs motionless while keepers were preparing her for care. Our only temporary setbacks in this training procedure came when staff shortages caused cancellation of sessions; the elephant was always ready to make progress. Fortunately, Portland was blessed with one of the most practically designed and complete facilities for elephant maintenance and separation in the world. Under the leadership of the zoo veterinarian, Dr. Michael Schmidt, many of the most advanced elephant reproductive studies are being accomplished in this facility. There is hope that zoos that are unequipped to properly house the more dangerous bulls will soon be able to receive elephant offspring as a function of new insemination procedures. In spite of the fact that this excellent attention is being paid to the elephants, there remains a majority of time in which there are few intellectually stimulating activities for these marvelous animals.

GENERALITIES ABOUT ELEPHANTS

During the time that we were conducting light-dark discrimination tests with a number of species, it was suggested by some of the zoo staff that our elephants could use some diversion and would be able subjects. Leslie Squier had worked with these same elephants some years before and tested their abilities to perform operant tasks. We knew that he had abandoned his work for a number of understandable reasons. For example, Les told us that during some testing sessions, keepers would literally come in and feed the animals while he was trying to work with them. The kindest thing that one can say is that the zoo was not terribly sympathetic to research during that period. However, in spite of such problems, Squier did manage to train three female Asian elephants to a criterion of 20 successive correct responses on a light-dark discrimination.

We searched around the zoo and found the remnants of Les's equipment on a scrap heap. It took considerable effort to refurbish the apparatus (Figure 9-1). The test apparatus measured 2.08 meters high by 1.24 meters deep by 0.86 meters wide and was made of plywood

Figure 9-1. Experimenters watch elephant's *(Elephas maximus)* response to discrimination apparatus. *(Photo by B. McCabe)*

with a slate operant panel. The response panel, set at a 25-degree angle on the front, was 0.79 meters wide by 0.91 meters high. Two translucent, round Plexiglas disks, 0.15 meters in diameter and separated by 0.36 meters, were centered 0.76 meters from the base of the panel. Two identical disks were spaced 0.51 meters below but were not used for this research. A galvanized steel feeder 0.15 meters in diameter was located in the center of the panel, 0.43 meters below the operable disks and 0.53 meters from the floor.

By reaching with its trunk, the elephant pushed a disk, triggering one to three microswitches located on the reverse side of the apparatus. Correct and incorrect responses were recorded on counters situated inside the box. The three-microswitch arrangement for each disk provided a uniform recording even when only one corner of the disk was pressed. Control and recording equipment was located inside the support structure of the operant panel that served as an enclosure for the experimenter. Floodlights (150 watts) mounted on swivel bases were adjusted to provide equal luminosity for each disk. Sugar cubes were delivered through a Plexiglas tube into the feeder that was constantly accessible to the elephant. We knew that elephants would work readily for sugar cubes that were not part of their daily diet, so no deprivation was required. (Squier had once discovered that sugar cubes were such a preferred reinforcer for elephants that they would literally work for them at rates where their metabolic needs could not possibly be met.)

Reinforcement was delivered for each response to the lighted side. The side correct was randomized with an intertrial interval of six seconds using a noncorrection procedure. With Dr. Squier's support and help, we were able to test the retention of these elephants in a formal task that they had not seen for more than eight years.

Because all of us at the zoo were excited about this unique chance, the word leaked out to the media. Imagine my surprise to arrive in time for testing the first of our elephants and finding television floodlights and flashbulbs in abundance. I told the reporters in the kindest way I could muster that it did not make much sense to test an animal's visual memory when the task was flooded with lights brighter than the stimuli. At the same time, I suggested that after the critical stages of testing were accomplished, it would be fine for them to take whatever pictures they wanted by simulating the original testing. Unfortunately, one person was quick enough on her feet to ask me how long that might be, and I had to confess that it might be minutes or months. One of our elephants, Tuy Hoa (born 1955 in Saigon), managed to save the day.

Those readers who have worked with animals in testing situations will recognize that it takes a certain amount of time for the animal to approach the apparatus, to learn how to respond, and to settle down to work. With the passage of eight years, we expected some delay in reacquisition. Instead, Tuy Hoa walked directly up to the apparatus as soon as it was in place, trunk-length outside her quarters. With no coaxing, she began to respond by pressing the lighted panels, making only two errors and taking a total of six minutes to reach 20 consecutive correct responses on this problem.

In some senses, this is obviously a remarkable performance and one that many *Homo sapiens* might have difficulty emulating. It saved our skin with the reporters and photographers who got their story with a minimum wait. On the other hand, it serves as another bit of evidence that these particular elephants had not lived a life conducive to gathering much potentially interfering information. Stated more simply, except for where she was expected to move or find her food, Tuy Hoa had not been asked to remember much else for the last eight years. Armed with this after-the-fact knowledge, we were prepared to find similar results for Rosie (born 1949 in Thailand) and Belle (born in 1952 in Thailand). It is fortunate that they were not the first animals tested.

For the next several days, we attempted to get Rosie and Belle to respond to the apparatus. Ultimately, after 1240 trials and 3 hours and 25 minutes of testing, Rosie reached criterion. She needed considerable help along the way, including some manual guidance to find the disks. Belle had an even tougher time. It took more than 647 hand-guided trials to get her regularly responding to the stimulus lights and a total of 2863 trials to criterion and almost 12 solid hours of testing.

It took a while for me to recognize that a large part of their successful responding occurred progressively in testing periods where the hand I used for guidance smelled of elephant mucous and tasted like sugar. Roger Henneous (the head elephant keeper) and I discussed the possibility that Rosie and Belle had some visual problems, a hypothesis generated in consultation with our veterinarian, Mike Schmidt. Although these big gals were getting along well in their daily routine, it was, as we have mentioned above, not terribly demanding. As I put it to Roger, my visually handicapped friends got around fine in their houses until someone moved the furniture, and no one had significantly altered the elephants' home environment. Consultants on vision were called in and retinal photographs indicated a significant vascular deficiency in the retinae of Rosie and Belle as compared with Tuy Hoa.

This serves to illustrate one of the most important fringe benefits of active behavioral research programs in the zoo. A significant physical anomaly was identified that might otherwise have gone undetected indefinitely. This, along with several other examples discussed throughout the text, provides evidence that even the simplest routine testing may help to provide signs of discomfort or deterioration in animals who are not equipped to report their symptoms in ways that we understand. This matter will be discussed in more detail in Chapter 14.

THE RUB-A-DUB ELEPHANT SHOWER

Working in Portland with access to the largest breeding herd of captive Asian elephants in the United States meant that considerable community as well as research interest was provided these remarkable residents. Even though the facilities have been described to be among the best existing with respect to zoo husbandry requirements, they do have remarkably limited space and variety of behavioral opportunity compared to the elephant's natural environment. Knowing that elephants frequently bathe in the wild and that they made wonderful use of a pond in their enclosure when it was filled, the zoo approached a local car wash manufacturer for assistance. The kindness, industry, and interest that employees of this company showed rivaled that of the local obstetricians who abandoned their patients to come help with our orangutan's breech delivery! They installed plumbing, sprays, water heaters, and necessary hardware around one of the major doors in the elephant enclosure (Figure 9-2). They also helped in the installation of a ring and chain that the elephants could use to turn on their own shower.

We intentionally did minimal training and simply demonstrated the shower operation to some of the elephants. When the chain was pulled, the shower sprayed tepid water for approximately 20 seconds. Each week, the elephants had several one-hour sessions in which they were given opportunity to activate the shower. Typically, three or four elephants, depending upon current husbandry considerations, had access to the shower, and all of them learned to use it by observing the few to whom we had given demonstrations. This installation took place near the end of our work in Portland, but it was instrumental in encouraging other zoos to provide similar apparatus. Although the elephants do use these showers with some regularity, we were at first surprised that the frequency of bathing was not considerably higher. Chapter 12 may shed some light on husbandry reasons that water may not have the universal appeal for captive elephants that it does for wild elephants.

Although this device worked acceptably in Portland, there are a number of caveats that should be issued. One is that in circumstances where water may collect, especially on concrete floors, perpetual wetness conditions may represent a potential health hazard, and elephants must be monitored for any related symptoms. Additionally,

Figure 9-2. Elephant pulls ring to start shower that surrounds door in background. *(Photo by G. Stryker)*

caution must be employed in monitoring water delivery systems in order to prevent excessive temperatures. The elephant has surprisingly sensitive skin. Finally, it bears repeating that whether one is attempting to enrich their lives with showers, new testing procedures, or ways in which they can do some healthful work, the tremendous power of the elephant must be respected. Within the time that I have written this book, two long-experienced and excellent elephant workers have been killed while working with their animals. In both cases, there is every reason to believe that the deaths were results of accidents, and in one case, the tragedy was extended by the fact that the elephant attempted to revive and protect its unintentional victim.

But the beauty and the wonder of the elephant continue to make work with it attractive in spite of occupational hazards. Here is a creature awesome in size, gifted in intellect, willing to do enormous amounts of work on behalf of human companions, and that has, sadly, an ever-decreasing habitat in the wild. A great challenge exists for the design of naturalistic exhibits that will encourage the elephants to use their remarkable prowess in ways that educate visitors about their capabilities.

10
SPECIAL DESIGN CONSIDERATIONS
FOR THE RAIN FOREST

This chapter will describe in detail two of the projects that were accomplished for the new Panaewa Rain Forest Zoo in Hilo, Hawaii. Although specific electronic components will not be identified, block diagrams and photographs of the apparatus may help the reader to understand in detail the ways in which we planned and accomplished this work. Jim Juvik, who was planning and supervising the development of this zoo, invited us at an early stage to incorporate some naturalistic behavioral engineering elements. This was really an exciting adventure! Although we had a somewhat limited budget, there was reasonable carte blanche to develop plans for species of our choice. It was also decided that the students in our annual summer course would take part in both baseline observations and the initial training of the animals to use the behavioral devices. Both Juvik and the architects were especially kind in providing electrical and structural requirements in their plans so that our apparatus could easily be accommodated.

Two of the major projects which were chosen were a gibbon brachiation apparatus and some field activities, including artificial prey capture, for tigers. Both of these projects were incorporated in an acre-size Sumatran swamp exhibit that was lush with vegetation, including both a bamboo forest and evergreen trees (Figure 10-1).

GIBBON APPARATUS

An artificial island for gibbons was surrounded by a large concrete-based pond. Rising from the gibbon island was a massive pole struc-

Figure 10-1. Bamboo forest is part of Panaewa Sumatran swamp exhibit. *(Photo by H. Markowitz)*

ture. A stainless steel cable connected this pole structure with the gibbons' hut which was 28 meters away (Figure 10-2).

We had a number of conflicting motives in designing behavioral opportunities for the gibbons. First, we wanted to be able to allow them as much freedom as possible within the exhibit, but we knew that even within a one-acre enclosure, there would be considerable danger that some gibbons might be lost to predation by the tigers. Tigers are one of the cats that have abundant aquatic abilities, and it

Figure 10-2. Panaewa gibbon island with cable to hut. *(Photo by H. Markowitz)*

was decided that initially the gibbon activities would have to be limited primarily to the upper levels of the pole structure (Figure 10-3).

The second major consideration had to do with the mode of feeding. Ideally, we would have preferred to allow the gibbons to gather fruit on the run in the foraging manner which constitutes a majority of their wild food gathering (Chivers, 1972). However, we also needed some way to guarantee that the gibbons could be easily collected for routine health examinations. In most zoo facilities, gibbons are very difficult to capture. Consequently, some zoos only deliver health care in emergency situations. This is unfortunate because, as with most primates, gibbons are susceptible to all sorts of human-borne diseases that they are exposed to in captivity. We decided on a compromise to encourage active behaviors in the collection of food, but

Figure 10-3. Tigers *(Panthera tigris)* romp in water in Sumatran swamp exhibit at Panaewa Zoo. *(Photo by H. Markowitz)*

provide the actual feeding in the home hut. This allowed ease of restraint by simply closing the door behind the gibbons on those occasions when health care or examination was necessary.

The stainless steel cable was wrapped with hemp to give the gibbons a more comfortable and firm handhold and to provide a more naturalistic appearance. A half-dozen rope vines were suspended from the pole structure. Inside the vine hangers, which were designed to appear as natural as possible, we placed special sealed mechanisms to detect when each was used. We decided not to require any particular sequence of behavior by the gibbons. Instead, the apparatus would simply count the number of vines which were swung upon and deliver food in the home hut in proportion to this random activity. The block diagram in Figure 10-4 illustrates the major components of this simple paradigm.

The controls available to the zoo staff allowed a selection of any number from one through six. Thus, if the number three was selected, it meant that movement between any three different vines would deliver a piece of fruit to the gibbons' hut. There was no contingency requiring immediate collection of this food. Instead, if the animals wished, they could swing around for an extended period and then collect a batch of food at home. Or, one animal might choose to exercise while others collected the fruit.

The recording apparatus automatically indicated the number of pieces of food that had been earned, the number of times that each of the vines had been used, and which vines were currently being counted toward criterion (Figure 10-5). Besides accumulating useful research data, these counters and data annunciators provided an indication of any possible malfunctions of the equipment. The keepers also knew exactly how much the animals had fed themselves and whether there was any need for supplementary feeding.

Although this is a rather simple design, it can be used to illustrate the preparation needed to allow for alternative ways that the residents might choose to use the equipment. If the family of gibbons increases in size and all are active on the island apparatus, the staff can simply increase the ratio requirement to guarantee continued activity without an excess of food. Should individual gibbons choose only to be

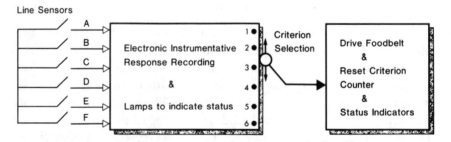

Figure 10-4. Major components of the gibbon apparatus designed for Panaewa. *(Diagram by Diane Fenster)*

Figure 10-5. Controls and data devices for Panaewa gibbon apparatus. *(Photo by R. Fial)*

food eaters and never food gatherers, alterations in the location of food delivery may be necessary. The apparatus is intentionally designed to facilitate such changes. Each of the components is modular in the sense that the food delivery belt can be moved to any area of the gibbon house and additional response requirement locations can also easily be established as add-ons.

Testing the Gibbon Apparatus

Unfulfilled promises about delivery times have become an expected part of everyday modern life. But these problems seem particularly magnified in zoos. After all of our installation work had been accomplished and the summer students had arrived in Hilo, the gibbons, which were now a year overdue, had still not arrived. So, out of urgency, we decided to use a surrogate gibbon in the form of a spider monkey *(Ateles geoffrey)*. Figure 10-6 shows Louise moving between the island and the hut. In a short time, she gained considerable ability in collecting food and occasionally spent her spare time tantalizing the tigers by hanging down and watching them prance around in the water beneath her. Despite the fact that spider monkeys have a tail to serve as an extra brachiation mechanism, their swinging and leaping is not as spectacular as that of gibbons. But, Louise's work on this island was attractive to visitors and to the staff, and we knew that when the gibbons did arrive, things would work well.

TIGER ACTIVITIES

Although as recently as 1936, the Sumatran tiger *(Panthera tigris sumatrae)* was still numerous and not considered in need of protec-

Figure 10-6. Spider monkey *(Ateles geoffrey)* moves between island and gibbon hut at Panaewa Zoo. *(Photo by H. Markowitz)*

tion (Mountfort, 1973), within the four and one-half decades since that time, the number of Sumatran tigers has dwindled to a few hundred. Its range is now limited to the northern part and mountainous southeast region of the big island of Sumatra. Much of its habitat has been eliminated by human encroachment, and many of the tiger's natural prey have also been destroyed. The tiger often has little choice but to prey on domestic livestock if it is to survive outside captivity. This predation has in turn led to continued hunting and destruction of the Sumatran tiger by local residents in spite of recent laws enacted in an attempt to save the species.

It is conceivable that within most readers' lifetimes, the only opportunity to see this smallest race of tigers may be in visits to zoos and other wildlife reserves. Unfortunately, most facilities will not be able to provide large game as potential natural prey for the tigers in their care. In the wild, tigers may spend a number of days between kills if the prey is substantial in size, but they are extremely adaptable and with smaller game may eat rather frequently. When food is scarce in nature, tigers have been reported to eat such unexpected tidbits as bird eggs, frogs, snakes, fish, and termites, and, although they are primarily carnivorous, they consume small proportions of berries, leaves, and grass.

In Panaewa we were limited to feeding butchered flesh from cattle and horses as the major part of the tigers' diet, but we decided to design a means by which they might experience some of the healthful exercise that comes with stalking and chasing two of their regular food items: rabbits and squirrels. This was an ambitious project, combining ideas for dynamic graphics and public participation with field exercise for a species for which it was impractical to provide natural prey. Before describing the individual components, it may be helpful to run through the entire completed sequence from the visitor's viewpoint.

Visitors entering the pavillion that overlooks the major part of the Sumatran swamp, may observe a television screen that is scrolling out information (Figure 10-7). Most of the time, this information concerns the differences between this exhibit and the real Sumatran swamp, some facts about the diminishing population of tigers in Malaysia, and some general ideas about the way that tigers prey on game in the wild. Should the tiger decide it is time to exercise and eat while the visitor is reading this information, there is an abrupt change in

Figure 10-7. Control console with public video display describing tiger activities at Panaewa. *(Photo by H. Markowitz)*

the television screen indicating this interest. In the wild, one of the ways in which male tigers loosely mark territories to defend and females mark hunting areas, is by scratching trees. Consequently, we equipped one of the trees with a special sensing device (Figure 10-8) so that scratches could be differentially detected.

Besides telling the public that the tiger has scratched its favorite tree indicating an interest in "hunting," the television screen invites the public to participate by randomizing the order in which activities will become available. If the visitor does not choose to participate within a half-minute, the computer automatically randomizes the occurrence of activities. Two of the animals' opportunities involve automated natural-appearing ground prey (Figures 10-9 and 10-10). A squirrel and a rabbit each run across four-foot mounds. When the tiger swats them or pounces on them, this represents a capture. The tiger has relatively unlimited time to make the capture and throughout this period, the screen continues to enunciate, for example, "The

Figure 10-8. Tree conceals scratch detection device that tigers use to initiate hunt. *(Photo by H. Markowitz)*

rabbit is running;" "The rabbit is captured." A six-foot berm is centered between these widely spaced ground prey (Figure 10-11). Tigers are great jumpers, able to leap more than 15 feet vertically. We reasoned that a little climbing and jumping exercise would be a naturally interesting behavior. When the tiger reaches the top of the berm, a treadle detects its presence and signals the computer which relays the message to the public that this activity requirement has been satisfied.

After all of the activities have been accomplished in the random order selected by the visitors or the computer, fresh meat is automatically delivered from a refrigerated area beneath the pavillion in which

Figure 10-9. Tiger *(Panthera tigris)* makes first capture of artificial prey in Pana-ewa Sumatran swamp exhibit. *(Photo by H. Markowitz)*

the visitor is standing. The final observable behavior is of the tiger rushing toward the visitor's pavillion and then disappearing from sight. In the meantime, the television screen tells the visitors that the work is completed and the animal is eating. Shortly, the screen begins once again to scroll information about the tiger's natural home and behaviors. This continues until one of the tigers next decides to initiate a sequence by scratching the tree.

As can be seen from Figure 10-12, there is much besides capturing these artificial prey and exercising on the berm to occupy the tigers'

Figure 10-10. Completed ground prey hill blends with earth (in right foreground) as part of Sumatran swamp exhibit. *(Photo by H. Markowitz)*

time. They romp in the bamboo forest, play in the pond, stalk each other, and occasionally try to capture birds or the resident primates. One of the most attractive things about this exhibit for those of us who designed it was that the tigers might space out their activities and respond in a variety of ways. The first time that we turned on one of the ground animals and the tiger began to look at it in the distance and stalk it (Figure 10-13), we were delighted. We were also

Figure 10-11. Tiger stands on detector plate at top of berm. *(Photo by H. Markowitz)*

very happy to see that the tigers did not spend all of their time in the area of the food delivery apparatus, but instead returned to it at unpredictable intervals.

The block diagram in Figure 10-14 includes the major components of the tiger system. Installation of these components was tedious, but most of us kept our humor. Juvik chipped in and helped bury the ground animals that weighed more than a half-ton (Figure 10-15), and we all crawled around in the mud of the swamp installing the automation and the artificial hills. (They looked very much like Martian pods on the way out to installation—see Figure 10-16.)

One special feature of the artificial ground prey was that they had to be very tough to withstand the punishment of a several-hundred-

Figure 10-12. Tigers play at edge of bamboo in Sumatran swamp exhibit. *(Photo by H. Markowitz)*

pound tiger and, simultaneously, they could not represent a danger to the predator. This was accomplished by a system of counterbalances that guaranteed that, regardless of the manner in which the tiger pounced upon the squirrel or rabbit, the prey would quickly fall into the hole in the hill without the possibility of catching flesh or claws. The covering upon which the final cosmetic layer of earth and epoxy was placed was welded from heavy steel components to preclude the possibility of the animals being injured by collapsing parts of the equipment.

One last note of interest has to do with the prior experience of the tigers who came to live in this exhibit. During initial training, we were surprised that even when they were hungry, the tigers would not accept meat that was thrown on the ground. The mystery was cleared up when we found that these animals had never before eaten off any

Figure 10-13. Alert tiger begins to stalk artificial prey. *(Photo by S. Rosenbaum)*

surface that was not concrete or asphalt. Placing a tiny slab of concrete where the food was delivered led to immediate acceptance of the food by the same tigers. This illustrates well the point that one cannot always count on captive animals learning to behave in the way that the field literature suggests they should.

SOME FINAL THOUGHTS ABOUT PANAEWA

In spite of the steamy heat of the Hilo summer, the class accomplished a great deal, and we recorded some unforgettable experiences. Among the apparatus that we built for the Panaewa Zoo was the speed game for the spider monkey described earlier. Shortly after the class ended, the Mayor of Hilo wrote to us with many kind words about the summer projects and an admonition for having built a game at which

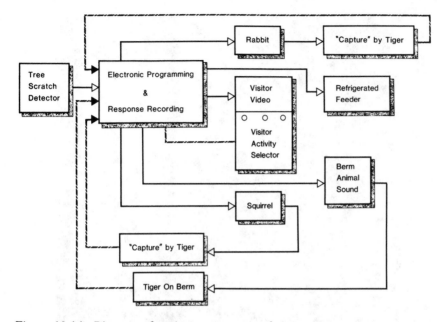

Figure 10-14. Diagram of major components of tiger activities in Panaewa Sumatran swamp exhibit. *(Diagram by Diane Fenster)*

a monkey could beat him. Faculty members at the University of Hawaii continued to work with the projects, but, unfortunately, staffing at the Rain Forest Zoo has not been sufficient to ensure everyday animal training protocol. As I write this chapter, I have just received word that, after a two-year delay, a permanent zoo director has finally been appointed and has already begun to work with the equipment. I hope to return to Hilo in the near future to continue the work described above and to do some planning for possible implementation in the other major exhibit, the South American rain forest.

This rain forest (Figure 10-17) has the interesting feature that the public walks in a cage that meanders through the center of the enclosure. Thus, species which might otherwise prey upon each other, are divided. This gives the public the impression that they are in a mixed exhibit. The lush vegetation which we described for the Sumatran swamp is also apparent in this exhibit, and our proposals for future work involve ways to provide tree-feeding experiences for the squirrel monkeys *(Saimiri sciureus)* without requiring constant refor-

Figure 10-15a. Artificial prey apparatus prior to installation. *(Photo by H. Markowitz)*

estation. We also designed some potential ways to encourage ground-burrowing activities for animals such as the patagonian cavy *(Cavia aperea)*. For the monkeys, portable radio-controlled systems that could be moved to various stout tree limbs, would serve as stations to detect movement. In a fashion very similar to the one described above for gibbons swinging on vines, movement between these tree limbs would allow the squirrel monkeys to accumulate food in an area where the keepers and veterinary staff could provide them necessary care. The rationale for making the devices easily movable from tree to tree was so that we could both provide varied activity and increase movement in those trees which could best stand the monkeys' current use.

Figure 10-15b. Automated rabbit being installed in ground enclosure. *(Photo by R. Fial)*

The ground-burrowing equipment would be attached directly to the keeper passageway in the center of the exhibit for ease of service. A food belt, connected through a mechanical linkage to an underground burrowing wheel, constitutes the major mechanism for this enrichment device. Burrowing by the cavies would move this wheel which would directly deliver them food. Initial on-site planning allowed us to select positions for the equipment that ensured visitors a maximum chance to watch this species-typical activity.

A number of us in the Research Center also designed and constructed a teaching machine for Panaewa Zoo visitors, similar to ones which we had built for the Portland Zoo. This machine used large projections of 35-millimeter slides with questions superimposed on them to invite visitors to learn about animals. In a typical presentation, a picture of a cavy might have superimposed upon it four identifications, only one of which was correct. Items would not advance

Figure 10-16. Four foot hills constructed of steel, concrete, and epoxy mixed with earth prepared for installation in Sumatran swamp exhibit. *(Photo by J. Juvik)*

until the visitor correctly identified the animal in the color photo. This format has proven to be much more attractive, based upon the total number of uses by the public, than traditional noninteractive graphic systems.

In one similar machine, which we built for use in Oregon, sound was also incorporated and children were encouraged to play in "Sesame Street" language. Each time that they made an incorrect answer, a differently phrased message would encourage them to continue, and they were verbally applauded when finally successful. Although this machine required considerably greater maintenance than the silent models, it did have one major bonus beyond its special attractiveness to children. This was the fact that preverbal youngsters were encouraged to learn to read the names of the animals which were shown on the slides at the same time as they heard them read aloud. The major maintenance requirement involved the magnetic tape assembly which was used to provide voice. Today it would be

Figure 10-17. Students are encaged in center of South American rain forest exhibit in Panaewa. *(Photo by H. Markowitz)*

plausible to construct a relatively maintenance-free machine using electronic voice simulation. Perhaps someday soon some of my students and I will have a chance to build such a machine for the kind people on the Big Island.

11
EXHIBITS FOR A THEME PARK

Marine World-Africa USA (hereafter abbreviated Marine World), a relatively young institution in Redwood City, California, has supplied research opportunities for a number of my students. A major concentration in this park involves "shows" featuring dolphins, whales, big cats, chimpanzees, and so forth. Visitors are encouraged to come in close proximity with some of the animals in hoof stock areas, and thousands of people each year take rides on some of the elephants and camels. Other more delicate or dangerous specimens may be observed from a "river raft" ride that winds around the perimeter of the exhibits.

There are difficult philosophical questions about the merits of exhibiting animals in shows involving artificial themes and inviting people to ride on resident creatures. At one extreme, it may be argued that excessive contact, stereotypical training procedures, and the inevitable "pollution" that comes on weekends when more than 10,000 visitors per day may attend produces unnecessary hardships for the animals (Batten, 1976). Arguments favoring this approach to animal exhibition include the facts that exercise is beneficial, and that the captive animals serve as "ambassadors" for their species and teach visitors who may know nothing at all about complex animal capabilities that creatures other than *Homo sapiens* are versatile and intelligent. Obviously, a complete consideration of the validity of this approach would require an entire textbook.

There were a number of things that attracted us to do some work in this park, beyond the kindness with which we were welcomed. Most important was the fact that management, trainers, and handlers universally accepted the principle that the animals' health came first. This was in marked contrast to some show parks, circuses, and carnivals where the "show going on" is interpreted to mean that even sick or otherwise endangered performers must work. We also found workers at all levels in Marine World open to the idea that it would be preferable to allow the animals as much flexibility and control of their own environments as possible in the work which we planned. It is my hope that someday parks such as this will have an increasing emphasis on naturalistic themes and on allowing the animals to provide much of the "script."

OTTER INSECT-HUNTING

Small-clawed Asian river otters *(Amblonyx cinerea)* are not used in shows at Marine World, but are exhibited in an area that includes an artificial pond, rockwork, and a small waterfall. Interestingly, the otters managed to dig some additional entries into the artificial rock beyond those originally planned by the builders.

Amblonyx are the smallest otters and reach typical weights of less than 12 pounds when fully grown. Their remarkable specially adapted, hand-like front paws are similar to those of the African clawless otter, *Aonyx.* (For years these two otters were both classified in the *Aonyx* genus, and some specialists prefer that taxonomy today.) The small-clawed otter has crushing molars similar to those of the sea otter *(Enhydra lutris)* which are appropriate for a diet including shelled crustaceans. They commonly manipulate objects in their forepaws and use their considerable dexterity in vigorously investigating all of the holes and crevices in their environment. Although there is little field data concerning the diet of *Amblonyx,* it is reasonable to infer that it includes insects, crabs, and a variety of other crustaceans that have been shown to constitute a major portion of the *Aonyx* diet.

Pat Foster-Turley and I decided to test the Marine World otters' interest in capturing insects (Figure 11-1). It was a real education to see the variety of responses during this initial testing. One of the otters reached its hand into the small opening provided and pulled out one cricket at a time, chomping them down as quickly as possible. Another preferred to play with them, watching them jump and occasionally putting one in the water to watch it "swim" before consuming

Figure 11-1. Small-clawed Asian otter reaches for live insects in coffee can. *(Photo by P. Foster-Turley)*

it. With encouragement and a small budget from the park's management, we developed apparatus to teach the visitors a little about otters, to give the otters a chance for an active insect hunt, and to provide some of the natural motivation that can best be accomplished by using live prey. In other parts of this book, arguments for exhibiting predator-prey interactions have been presented. It is interesting to note that in this case the administration was quite apprehensive about using larger prey species, including fish. Insects, however, were a live food upon which we could agree. In fairness, this commercial park has serious concerns about any adverse visitor reaction, and it is only natural that they should prefer experimental projects that minimize complaints. The unfortunate fact that most people see insects as things to be squashed reduces concern about using them as live prey.

Graphics were designed to introduce visitors to the otters' new opportunities and to invite their limited participation. The sign inviting interaction from the public included a picture of the major hunting areas for the otters and buttons with which visitors could select the area in which prey could be successfully captured (Figure 11-2). The same panel also included electronic counters that displayed the number of hunting opportunities in each area for the current day. A major graphic with four separately illuminable sections annunciated the entire sequence of events (Figure 11-3).

Master control and recording components were manufactured with integrated circuits, and the electromechanical peripheral devices used were constructed in the shops at San Francisco State University and at Marine World. In a final form, they were assembled to produce the following events. At the beginning of the day, the sign saying "Watch otters hunt" invited public participation. If no one chose the area where crickets could be obtained, the apparatus produced a hunting opportunity on a random basis after a preselected interval. In either case, a cricket sound became audible to otters and visitors. This sound was generally broadcast throughout the area and did not provide a specific clue concerning the location of the crickets. Spectators were told that the otters used the sound to tell that they might hunt if they wished to and that if the otters chose not to hunt now, they would have additional opportunities later.

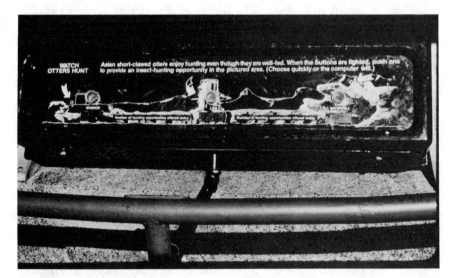

Figure 11-2. Graphic keyboard allows Marine World visitors to participate in selection of prey location. *(Photo by P. Foster-Turley)*

Figure 11-3. Master graphic describes hunting opportunity for otters at Marine World. *(Photo by P. Foster-Turley)*

User-adjustable instrumentation allowed selection of maximum intervals for hunting to begin. There were three hunting areas in the exhibit: one close to the visitors, one behind a tree stump, and one above some major rockwork. The otters' entry into the insect capture spaces within these hunting areas was automatically detected with photoelectric equipment. As long as entry was made into any of these areas, hunting was allowed to continue. When the otters entered the section where insects could be captured, the cricket sound ended. This served to indicate to the otters that they had successfully completed the hunt, and it served as a "bridge" to let them know that their prey was forthcoming.

Three identical live cricket dispensers were fashioned for the three hunting areas. Each had 50 compartments (Figure 11-4) that could store as many as several crickets, depending upon their size. When a successful hunt was completed, the instrumentation signaled the appropriate feeder that advanced 7.2 degrees and dropped the next group of crickets into the appropriate hunting space. The second segment of the major graphic was then illuminated, indicating that insects had been caught. In cases where the timer expired without active hunting, an illuminated graphic said: "There's more to life than hunting." A rest period followed, and the final graphics segment de-

Figure 11-4. Crickets in dispenser at Marine World. *(Photo by P. Foster-Turley)*

scribed cooperation between the park and the University in providing naturalistic opportunities for the otters to exercise their hunting skills.

Throughout the design and construction of this equipment, we tried to keep the apparatus as universal as possible. Thus, prey items could easily be changed to other insects without altering mechanical or electronic components. Eighteen electromechanical counters were used to total all of the pertinent information from the day's work (Figure 11-5). This included responses by the public as well as the otters and entries into the area at times that did not lead to capture. Provision was also made for on-line recording of all identifiable events for research analysis. Fortunately, some of the very things that interested students focusing on research questions also were of interest to the park management. For example, all of us wanted to know how much each segment of the exhibit was used so that we could improve its design in future renovations. Similarly, there was uniform interest about the extent to which visitors appropriately interacted with the equipment as compared with haphazard pushing of the buttons. We knew from previous work that some people would push any button in sight, and at the time I was completing this book,

Figure 11-5. Final instrumentation and one of three feeders installed in otter exhibit at Marine World. *(Photo by P. Foster-Turley)*

we were looking forward to quantifying the proportion of useless responses in this particular exhibit.

Figure 11-6 is a general functional diagram illustrating the major components of the instrumentation. This drawing and the final version of the master panel were accomplished by John Noble and Jerry Ross of the San Francisco State University Science Service Center.

Prior to inviting public participation we ran many trials to ascertain the otters' response to this new opportunity. They busily hunted in the areas we had installed (Figure 11-7), and when they eventually captured crickets these were often treated with much more interest than their regular food (Figure 11-8). Equally important was the fact that the otters were able to ignore the signal that prey were available when they wished to engage in other activities (Figure 11-9).

Having attention to a hunting opportunity annunciated by a speaker system is not the most naturalistic possible situation, but the otters responded with very little coaxing, and, once they had grasped the

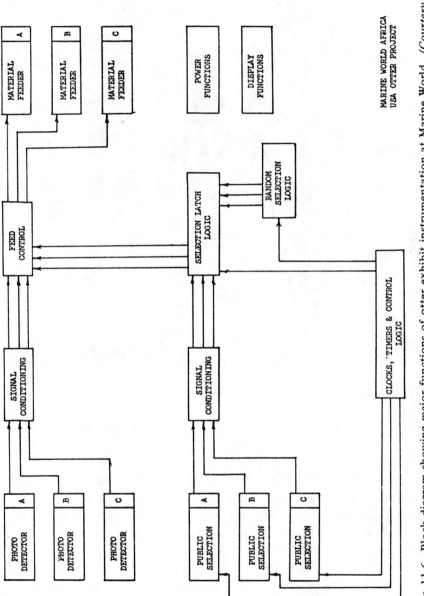

Figure 11-6. Block diagram showing major functions of otter exhibit instrumentation at Marine World. *(Courtesy of San Francisco State University Science Service Center)*

Figure 11-7. Small-clawed Asian river otter searches for insect prey. *(Photo by P. Foster-Turley)*

consequences of their investigation, they were tireless hunters. We did not deprive them of any part of their regular diet, and the crickets—which weigh very little—were seen as an added treat. The otters' behavior became increasingly efficient and natural in appearance as time progressed. When they heard the cricket sound and were interested in hunting, they would often return to the place of previous insect capture (just as an animal might in the wild). If no prey was found there, they would hunt in other locations, sometimes individually and sometimes in tandem with other otters.

During the time that we were working on this equipment, park employees who were our friends began a systematic campaign to tease us (and the park administration) about the use of live prey (Figure 11-10). One of the promoters of the "Save the Cricket Movement" said that her major motive was to have fun in the midst of an occa-

Figure 11-8. Otter douses prey that it has captured. *(Photo by P. Foster-Turley)*

sionally monotonous job. But a second motive was that this intentional *overemphasis* on "protection" of animals that would ordinarily be consumed after being killed might help to gently persuade folks in control that the benefits of using live prey in teaching visitors about natural behaviors might outweigh worries about negative public reaction. Whatever the primary motivation, these campaigners were a diligent lot and picketed us with "Save the Cricket" signs during the employee party to dedicate the exhibit and during the filming of a sample hunting session by a crew from Britain who was doing a show for BBC.

We hope that someday we will be able to combine the insect-hunting opportunities with chances to capture fish for these elegant small-clawed otters (Figure 11-11).

SOME EXERCISE FOR CAPYBARA

Capybara *(Hydrochoerus hydrochaeris),* largest of all the rodents, are avid swimmers that live in groups near lakes and rivers in tropical

Figure 11-9. Two otters ignore sound of available prey to nuzzle each other. *(Photo by H. Markowitz)*

America. Their display at Marine World was rather limited in size, but did include a large pond, some grassy areas, and a hut. These particular capybara did not voluntarily allow handlers to approach them and kept their distance from visitors who came to the fence surrounding their home. Despite the limited size of the exhibit, we decided to provide a variety of methods to encourage appropriate exercise and allow the animals to feed on their own demand. Research observations prior to the design of the equipment indicated significant aggression by the more dominant rodent when the others approached feeding areas; this reinforced our decision to provide a variety of ways in which animals could simultaneously feed. Two wooden paddles, fashioned from large oars, were suspended above the water level along the side and near the deep end of the pool. The capybara could contact these in whichever order they wished and, after both

Figure 11-10. Sign spoofs criticism of live prey approach to enrichment activities for otter. *(Photo by P. Foster-Turley)*

paddles had been moved, food was automatically delivered to the opposite side of the pool. This encouraged movement to most of the perimeter of the pool from the shallow area where they normally stayed.

The second set of apparatus we designed involved detecting the capybara as they approached a large post near one end of the exhibit and as they moved across a small hill at the other end. Once again, the rodents could make responses in whichever order they preferred, but movement in both areas was required to deliver food. This food, earned by movement on land, was to be delivered in a distinct location from that provided for aquatic activity. Both of the food-delivery mechanisms were composed of universal film loop feeder belts (described in earlier chapters) and a conduit that channeled the food to appropriate parts of the exhibit.

Figure 11-11. Small-clawed Asian river otter carries a meal of fish in its mouth. *(Photo by P. Foster-Turley)*

Since capybara also do considerable ground scratching and occasional burrowing, the final device we designed was used to simulate this behavior. A small treadmill was placed in an underground opening and was used to directly drive a feeder belt. The capybara could thus reach down below the surface of the exhibit, move their paws against the apparatus, and "dig away" to produce food.

With all three sets of apparatus installed, the opportunity was provided for each of the capybara to work independently and feed independently or for them to work simultaneously at the same tasks. Unfortunately, in the midst of this preparation, a management decision eliminated the display of this slow-moving species in favor of more active animals.

TIGER ISLAND ACTIVITIES

Visitors on the river rafts and the opposite shore can observe Bengal tigers *(Panthera tigris tigris),* often in direct contact with their trainers and handlers, in a section of Marine World called Tiger Island. Daily protocol involved about seven hours in which some of the tigers and handlers played, wrestled, and gave various demonstrations for park visitors. Some of the older, tougher, or more irascible tigers did not take active part in these activities because of the danger to the handlers. The first motivation for participation in this area was to provide active opportunities for those particular tigers.

The park had earlier spent considerable money installing a large slide area that led to a pond where the tigers liked to play. Although this slide was somewhat more naturalistic-looking than many in animal parks (Figure 11-12), the tigers did not actively use it, except for sunning themselves at the top. We discussed with the park staff their attempt to train the animals to use the slide and the difficulties that they had encountered. Essentially, they had managed by successive approximation to get the tigers to the top of the slide and then coax them over with a gentle push. From most of our own work with cats it seemed clear that the procedure outlined here was almost certain to result in increased wariness about the slide. With the active assistance and support of the tiger handlers, we planned equipment that would allow more gradual and patient encouragement for the tigers to use this part of the island.

Figure 11-12. Tiger-island slide at Marine World-Africa USA. *(Photo by D. Hearst)*

A portable device incorporating an audible stimulus, a tiger detection unit, and an automatic meat-chunk dispenser was designed. In order to minimize danger to the tigers and their handlers we wanted to avoid running any exposed cables, so the unit was battery-operated and radio-controlled. This should facilitate a very simple training procedure in which any of the tigers, including those difficult to handle, can be painlessly led in small increments to try using the slide. First, the animals will be taught that when they approach the apparatus when it is chirping, they will receive food. Then the training device will be gradually moved up the side of the slide that the tigers climb to encourage them to reach the top to be fed. Finally in this section of the shaping procedure, the apparatus will be gradually moved down the surface of the slide until the tiger has to commit itself to sliding in order to obtain the food. After each of the animals has learned to use the slide with greater comfort, a new apparatus will be installed so that the tigers may learn to move through several areas of the slide and the pond in order to obtain food when they wish. Photocells will detect their movement through these areas, and, to lend some variety, they may splash in the pond or slide first as long as they eventually move in both areas.

Although we wanted to help with the park's dilemma about slide usage, most of us were even more interested in watching the beauty of the tiger as it ran and bounded around the island. To encourage this activity, we planned three devices and a training procedure to teach the tigers about this new opportunity. There was a large dead tree laying on its side on Tiger Island (Figure 11-13) that the cats occasionally used to scratch and to stretch themselves. The first device will simply detect this already-prevalent behavior. At the left-hand rear corner of the exhibit, there is a large area of greenery and an electrical barrier preventing the tigers' entry. The most elaborate part of this apparatus is designed for installation in this area. It includes a system for randomly flying chunks of meat which can be captured by the tiger on its safe side of the electrical fence. The apparatus will allow the handlers to load several chunks of meat independently at the beginning of the day so that they will not have to enter the area while the tigers are capturing food. In another corner of the island, the tiger will be detected as it moves onto an elevated area.

Once the tigers learn to entertain themselves by capturing their flying "prey," additional exercise will be encouraged by making this

Figure 11-13. Dead tree that tigers use to stretch and scratch. *(Photo by D. Hearst)*

opportunity contingent upon their having used both the scratching tree area and the elevated portion of the exhibit. In the final stage of this work, an ultrasonic detector will be placed in various locations among trees on the island. A sound generator that produces bird calls to attract the tigers will be included in the same portable equipment. Movement to the area of this apparatus will be an additional way for the tigers to activate the "prey." So, from the visitor's viewpoint, the tiger may randomly move as it wishes between the elevation and the trees, after which it will have a chance at capturing some food.

A SHORT CHASE FOR THE LION

Lions *(Panthera leo)* have never been a major emphasis of our work, partly because their daytime behavior in zoos is more comparable with that in the wild than is the behavior of many animals upon which we have focused, namely, lions sleep a lot. Although they sometimes show great perseverance in wearing down quicker animals, they are not noted for great pursuit activities. In nature, they more typically hunt at night and depend on the element of surprise in capturing prey.

At Marine World, several lions are allowed to roam in a large fenced area during the day. The visitor's primary view of these lions comes once again from the boat trip. During the summer when the majority of visitors attend, it is hot in Redwood City, and the lions tend to aggregate in one area under shade where they are not easily seen. Although we were reluctant to insistently wake the lions from their rest, we did decide to help the park staff encourage occasional simple activity. A large simulated tree stump was designed so that it could not be toppled by the lions, but could be moved by handlers. Incorporated in this stump was a radio receiver, a radio transmitter, and a lion detector.

The idea was simple: boat drivers could be provided a sending unit so that as they approached the area, they could tell their passengers that the lion *might* come forward if it felt like eating. The tree stump was made portable so that it could be used in different parts of the exhibit to provide some variety for the lions. When the tree stump was signaled by the boat, a warbling sound went on for a brief period to let the lions know that coming to the stump would result in delivery of food many meters away at the back of their enclosure. When the lions felt like working, passengers on the raft would be treated not only to the appearance of the lion on its feet and near the boat, but they then might see it move swiftly toward the back of the cage to be the first one to obtain food. We are currently working on the installation of this equipment.

OTHER IDEAS IN THE PLANNING STAGES

Many of the animals in show parks are selected because of their complex capabilities and their high spontaneous rates of activity. As a matter of fact, many of those behaviors that visitors find the most spectacular or exciting are natural behaviors which have simply been brought under contingency control. Although I would personally wish for more species-appropriate biological content to be conveyed in shows—including those at Marine World—it seems clear to me that the activity provided is far preferable to the sterility of environments that provide no motivation for action. However, there are within show parks "retired" performers such as animals recuperating from illnesses. There are also some species, such as the Steller sea lion *(Eumetopias jubatus),* that are impressive to see but difficult to use for work requiring direct contact. A number of my students and co-workers are currently concentrating on ways to provide special opportunities in these cases.

In one holding area at Marine World, dolphins are kept in a large "petting pool" where the public is allowed to touch them under supervision, but is also warned that they may bite. There are many days in which these complex creatures keep some distance from most visitors and have little with which to occupy themselves. Several of us are currently thinking of varieties of games and other activities that these marine mammals might use to entertain themselves. Dolphins are so inventive that it seems certain to me that, given the opportunity to play with or invent flexible games of their own, they would entertain visitors at the same time that they amused themselves. It should be carefully noted that we are not proposing more "routines" for the animals to learn, but instead are suggesting a variety of activities to choose from if they wished to relieve the boredom of confinement.

One of the proposed new areas of the park would be a Steller sea lion exhibit to house these mammals that are currently in small pools off public view. Some of us are helping to plan active behavioral components that may be used for this exhibit. We hope to be able to exhibit the Stellers' great strength and surprising speed in ways not ordinarily seen in captivity.

This chapter may illustrate that people in show parks are indeed receptive to new ideas, willing to entertain experiments designed to provide naturalistic biological contingencies, and generally willing to provide carefully supervised research opportunities. It is my hope that these projects will help to encourage increasing proportions of exhibits where more varied opportunities are provided for animals when they wish to use them.

12

REAFFIRMING THAT ANIMALS
ARE SMARTER THAN
INVESTIGATORS

Once, in the early part of my graduate training, I designed what I thought would be an elegant shuttle maze for rats. This was an apparatus to test the effects of some drugs, and an animal had to run from one side to the other when a signal came on to avoid a mild shock. Because I didn't have very much money, I simply made a very high ceiling and covered it with hardware cloth. On the first trial, the rat demonstrated to me that nothing I had learned from books would be of much help in predicting animals' ingenuity and their proclivity to solve problems in ways other than those intended for them by researchers. The moment the shock came on, this rat jumped much higher than it was supposed to be able to and grabbed the hardware cloth. Then, proceeding in a fashion that I had always read was more appropriate for primates, the rat brachiated paw over paw to the other side and dropped down safely. I watched this in amazement for several trials, junked the apparatus, and started anew. But I do have to confess a sort of pleasant feeling about the rat's having been much too smart to abide my textbook-ish design. Through the years, it has become apparent to me that everyone has similar experiences, but few are foolish enough to talk about them publicly as I will here. This chapter will highlight a few of the odder ways that animals in zoos have redesigned the experiments with their own unique methodologies.

ON CAMELS AND RATIO STRAIN

Vic Stevens worked with us on the first formal studies of operant behavior in the camel *(Camelus dromedarius)*. A panel was installed
that allowed the camel to stick its head through an opening and respond by pressing lighted rectangles. This apparatus was situated to
minimize the probability of the camel's spitting on, or biting fingers
off, the experimenter. One of the first problems that Vic decided to
investigate in a preliminary way was the effect of increasing ratio
strain on performance for this species (Stevens, 1978). Adolph had a
tremendous rack of teeth and a prominent snout (Figure 12-1) that

Figure 12-1. Adolph *(Camelus dromedarius)* displays his teeth and nose. *(Photo
by H. Markowitz)*

we felt sure he would use to manipulate the key in much the same way that a rat pushes a key with its nose. The idea was to see if, by gradual increments, we could work Adolph up to high ratio requirements (e.g., 100 or more responses for a single reinforcer consisting of access to alfalfa pellets). Adolph solved the problem in his own idiosyncratic fashion. He stuck out his chin, learned to vibrate it at very high speeds and worked up to a ratio of 150 to 1 in a relatively short time.

We later modified the apparatus so it simply required touching rather than pressing, because there was some concern that Adolph might grind his incisors excessively with his selected mode of response. But, Adolph had reminded us all not to be so self-assured that we could predict a previously untested animal's responses, even to the most traditional-appearing equipment.

OSTRICHES WILL WORK FOR PEANUTS

Some of the best researchers who worked with us came from Reed College, and Wilfried Zimmerman, who later became our research coordinator, was no exception. His thesis was on ostriches, and the birds taught him a great deal along the way. One day, Zim came to me and described a peculiar side-to-side jumping behavior that the ostrich kept exhibiting, but only in Zim's presence. After we identified that this was a courtship behavior, Zim felt fortunate that he was using remote instrumentation rather than immediate contact methods with the ostriches.

An accidental contingency arose in the experimental work. Zim was doing some key pecking studies with the ostriches who lived in the same large open enclosure as some giraffes with which we were simultaneously working on visual discriminations. He found that the ostriches would work very hard when he used peanuts as reinforcers, but that they worked in unusual ways compared with other birds, such as pigeons for which there was an extensive operant literature. The ostriches did not take turns. Instead, they would all try to peck at the key at once and would often peck at each other in the process. There were no injuries, however, and the work was progressing well.

One day, one of the keepers did not notice that Zim was running his session and accidentally dumped a whole load of peanuts into the ostriches' trough at the same time that peanuts were being used as reinforcers. The ostriches went over and sampled some of the free peanuts that had been taken from the same batch as Zim was using, then returned and began responding on the key to earn the same food which they could have free.

At first Zim was angry about the intrusion, but soon we felt fortunate to have accidental corroboration of a significant contemporary laboratory result: assuming the effort required is not too great, many animals will continue to work for food which is simultaneously offered to them free in the immediate vicinity. As mentioned in the introductory chapter, this effect, first clearly described by Allen Neuringer (1969), has provided a lot of difficulty for many of the traditional reinforcement theories of learning. But it has now been so often replicated that few researchers deny the validity of this phenomenon. Certainly the ostriches, constantly competing with each other for access to the key, might have gone off and eaten the free peanuts. Instead, they opted to work for them. I leave it for others to speculate about the hypothetical explanation of this behavior. But the ostriches and the keeper accidentally taught us something we would not otherwise have learned for this species.

THE WALLAROOS THAT WOULDN'T JUMP

One of the bright young people who came to work in the program at Portland was Nancy Eldred, who had a great interest in marsupials and in particular in wallaroos *(Macropus robustus)*. A careful survey of the literature showed that, while experimenters had done a number of studies involving kangaroos, wallaroos, and wallabys, surprisingly no one had done an experimental analysis of their jumping. The keepers told us that in some cases they had literally seen the wallaroos jump from a standing position to a height above that of the fence, which at its lowest point was about seven to eight feet.

Nancy constructed some kennel club-type jumping barriers that were intended to be used to increase the jumping requirement by about six inches at a time. The plan was very simple. We would teach the wallaroos (roos) first to move back and forth between two widely spaced stations in order to obtain food. Then we would increasingly raise the barriers. The barriers were 18 feet wide, so if the strain became too much, the wallaroo could run "around end." It was reasoned that this would minimize the stress and give us an approximation of the wallaroo's jumping ability in a relatively voluntary situation. It looked great on paper.

Despite the fact that she worked diligently for a year, the major findings were: (1) nondeprived wallaroos didn't seem to care very much for any sort of reinforcement delivered by an experimenter, (2) when more proximal shaping methods were attempted, the emus who lived with the wallaroos proved talented at chasing the roos and the investigators, and (3) for a species with such a great reputation for general activity and jumping, the only effective captive stimulus seemed to be the sight of a keeper with a broomstick.

Nancy's research was also impeded by unpredictable zoo protocol, visitor intrusions, and so forth. Still, everyone on the scientific staff had a try at it, each one of us believing that the others must have blundered because we were sure *we* could get wallaroos to jump. Not so.

A YOUNG ELEPHANT OUTSMARTS AN OLD EXPERIMENTER

Although I greatly enjoyed working with all of our Asian elephants *(Elephas maximus)* and was particularly fond of one of our big cows, Tuy Hoa, there was a special place in my heart for young Gabriel. I came to love and admire him during one of the days when the Portland Zoo opened its doors to visiting scientists. The primatology meetings were held in town and, with the accompaniment of staff, the professionals from these meetings were allowed in the areas normally reserved for keepers and residents. Almost everyone was a considerate visitor, careful to observe protocol in the same way that all of us would have had we visited other laboratories or research stations. But, there was a significant exception. One graduate student, dragging his girlfriend behind him, stepped right across the chain that had been placed to separate the elephants from our special visitors.

There were large bars that separated the adult elephants from the major passageway that had been chained off. Gabriel was still young enough to walk between the most widely separated bars which had about 1.5 meters of clearance. As the student walked up to the bars and started to brag about his knowledge of elephants, Gabriel walked out and stepped on his sandaled foot. We were surprised, since ordinarily elephants are exceptionally careful *not* to step on unfamiliar things, and this response seemed quite deliberate. The friend asked, "Oh my god! Didn't that hurt?" and when her would-be hero said, "No," Gabriel stepped on it again! From that moment on, this elephant was one of *my* heroes.

Soon the zoo management decided to move Gabriel to the nursery area of the children's zoo, a move that stimulated me to design some experimental apparatus with which we hoped to test discrimination learning in a variety of young animals. I had the idea that it would be nice to build a universal apparatus using contact detectors (much like those used in modern elevators) so that a wide range of species could respond by contacting the panels in any fashion they chose. For example, I envisioned that snakes could sidle up alongside the detector and trigger it off, that elephants might choose to touch it with their trunk or rub against it with one of their legs, and so forth. The initial problem for Gabriel was to be very simple. All he had to do was to go and touch the large stainless steel disk underneath a lighted panel. There were two disks with associated lights and the correct side was to be randomized (Figure 12-2).

Gabriel's solution was more simple and ingenious than I would have liked. He always managed to store some water in his trunk, no matter how long before the experimental session we emptied his trough. Then when the apparatus was turned on, with great care he would smear the water between the two response panels, electrically shorting them so that a response on *either* side paid off. So, back to the drawing board we went to design a new multispecies apparatus with which to study discrimination learning. Gabriel certainly taught us a lot about the comparative abilities of *Elephas maximus* and *Homo sapiens*.

ON THE RELATIVE INTELLIGENCE OF ORANGUTANS
AND APPARATUS DESIGNERS

Each of the great apes has its own special temperament. Where the chimpanzee *(Pan troglodytes)* is famous for its mimicry, aggressiveness, and human-like behavior, and the gorilla *(Gorilla gorilla)* for its gentleness and private nature, the orangutan *(Pongo pygmaeus)* is often

Figure 12-2. Discrimination panel used with young elephant employs contact detectors to record responses. *(Photo by B. McCabe)*

considered by zoo workers to be the "engineer" of the apes. Given access to almost any apparatus, the orang will dismantle it in moments. During one period when we decided to show some motion pictures to the Portland orangs for their entertainment, Inji and Harry seemed most interested in the installation procedure. They watched me through the glass, tapping on the window, making motions about how to turn the screwdriver, and generally overseeing the project.

We knew that when we installed their discrimination apparatus, it would have to provide significant barriers against access to the equipment by the orangs. Not only do orangutans like to take things apart, but they are so strong that they seldom need tools. For example, when we had an emergency with one of Inji's offspring, we had to isolate her in a room which was not really meant for apes. Inji weighed only 96 pounds at the time and yet she was able to bend one of the water pipes off the wall. Imagine what Harry, who weighed 360 pounds could have accomplished!

The discrimination apparatus was carefully armored and mounted in place of a window in the steel door of the orang enclosure (Figure

12-3). All of the control and programming equipment was located safely in an anteroom. The reinforcement belt was placed on the side wall outside the cage with a chute designed to deliver the food into an opening in the orang's door. Our idea was that these apes would learn the reversal discrimination in the same way that we had tested many other animals (see Chapter 7). We proceeded by running trials in which responses to the panel which lighted were to be rewarded. Harry proceeded differently. With great care, the orang found one particular angle that allowed him to see the food belt in the adjacent room by peering through the reinforcement slot. Then Harry reached his great arm several feet up to the discrimination panel. Never looking at the lights which were supposed to represent the

Figure 12-3. Orangutan *(Pongo pygmaeus)* responds to discrimination task in home cage. *(Photo by R. Davis)*

problem, Harry watched the belt instead, apparently intrigued with its moving parts, and played the press panels like an organ, waiting to see which responses would eventually make the belt move. Some of our graduate students and technicians were dismayed at this turn of events, but I personally found it entertaining though perhaps not technically as informative as the discrimination results we obtained after closing Harry's peephole (Davis and Markowitz, 1978).

A GRIZZLY TALE

As training exercises for summer classes, we accomplished a few projects in the Honolulu Zoo. Jack Throp and his staff were kind enough to welcome us to work there and to scrape up parts from their small budget to help with the equipment. Although none of these projects were very long-lived, they did provide some interesting experiences, not the least of which came with the zoo's grizzly bear *(Ursus arctos horribilis)*.

When we first discussed the plight of the zoo with respect to this particular exhibit, two major problems were described. First, there was some concern that the grizzly bear was old, and, although he might profit from exercise, some of the staff thought the bear could not stand erect or even move any great distance.

The second problem was an especially interesting one. For years, visitors had thrown all kinds of foods to and at the grizzly. Jack loved the flourishing plant life in Hawaii and was not a great fan of environments that led to bears begging, so he had allowed the foliage to grow in front of the cage. This made it very difficult for people to get the necessary ballistic angle to fire food into the grotto without hitting the trees. They could still see the bear through a lovely set of greenery (Figure 12-4), and some of the bear's begging behavior slowly diminished. But, as might be anticipated, there were frequent complaints from visitors who felt disenfranchised from their interaction with the animal. On visits to the zoo, I heard people talk to youngsters about how as children they had fed the bear and how unfair it was that it was no longer allowed. We set about a project to return public feeding in a way that might reduce begging and guarantee nutritious foods.

A long conveyor belt (see Chapter 13) was assembled to drop food onto a catapult, and the public was invited to push a button at fre-

Figure 12-4. Foliage prevents visitors from throwing food into pool where bear bathes in Honolulu Zoo. *(Photo by H. Markowitz)*

quent regular intervals if they wished to see the bear receive some food that was good for it. The design of the catapult ensured that the food would fly to random areas in the grotto and precluded the bear from sitting in a single position to catch food. The idea was initially well-received by visitors and zoo workers, and the bear actually got up and moved around with some frequency. But, a few weeks after our class had ended and we returned home, my good friend Jerry Marr called to tell us about the evolution of the bear's behavior. Instead of moving around on a regular basis and exercising to get its food, the wise old bear simply waited for a number of visitors to catapult food into the grotto and then, with a minimum of effort, would gather it all at once.

Armed with this knowledge, we resolved to return the next year with some experimental solutions to this problem. In the meantime, what many of us had suspected from watching this grizzly during the summer—that it was neither exceptionally slow nor lacking in perception—was verified when it caught and consumed the kit fox that was living with it.

The following summer, two long manipulanda were built and installed near some rotating stimuli with pictures of prey animals. The new idea was to get the bear to move a significant distance across the back of the grotto. His responses would then enable the public to catapult food.

Before this project ended, we did learn two more things from the grizzly: those keepers who thought this bear incapable of standing erect were in error (Figure 12-5), and the long-entrenched learning of begging behaviors had become an almost indelible part of the bear's repertoire. When he wanted the apparatus to go on, he would occasionally beg it to do so by grabbing his rear paws with his front ones and spreading his legs while looking expectantly at the prey pictures (Figure 12-6).

TWO MORE ELEPHANTS THAT REMEMBERED

Another relatively short-lived project in Honolulu involved two summers of providing elephants an opportunity to shower in their very warm yard, which offered no shelter and little other relief from the heat. Some of the rationale for elephant showers has been discussed in Chapter 9. We were especially sure that Empress and Toto, the Honolulu Asian cows *(Elephas maximus),* would appreciate this break in their warm and monotonous routine. A massive shower chain was constructed on the roof of the barn so that it extended over the elephants' yard. Although some of the keepers told me that these elephants were occasionally unmanageable, I found them easy to work with, and, within the first week, Empress was willing to pull the chain. This training was accomplished by rewarding her with positive comments and with fruit, which I used to manually coax her trunk through the ring (Figure 12-7). Eventually she was willing, a good percentage of the time, to pull the ring for a few kind words and a slap on the trunk. But, alas, neither elephant seemed overjoyed at the

Figure 12-5. Grizzly bear *(Ursus arctos horribilis)* stands erect to respond to apparatus that catapults food into exhibit in Honolulu Zoo. *(Photo by H. Markowitz)*

prospect of a shower. As a matter of fact, they actively avoided the water, which was delivered from a hose on the roof of the barn.

Figure 12-6. Bear reverts to begging behavior "asking" the apparatus to throw food to him. *(Photo by H. Markowitz)*

Empress and Toto would occasionally go to the stagnant pool in one corner of the exhibit and bathe with this unsanitary water. We tried to account for this behavior with lots of alternative hypotheses. Perhaps they just liked dirty water—but when the pool was filled with fresh water, they were willing to use that; perhaps the shower-delivered water was too cold for them—we warmed it and it made no difference. All of these experimental attempts were to no avail. The real problem resided in the elephants' clear memories of a factor that most keepers had forgotten and none had remembered to share with

Figure 12-7. Empress, an Asian elephant *(Elephas maximus)* pulls ring to obtain fruit in Honolulu Zoo. *(Photo by H. Markowitz)*

us: the very same hose that we were using to "treat" the animals to refreshing water had, in years gone by, been used as a high-pressure force to compel them to move when they resisted the keeper's instructions. The elephants enjoyed pulling the chain for papayas, mangoes, bamboo shoots, apples, and even for a few kind words, but not for water from a source that had been used to push them around.

CONCLUSION

In general, one great benefit of giving animals increased behavioral opportunities is that they will discover unique ways to use them. For the laboratory researcher accustomed to eliminating all of the "controlled variables" as much as possible, the things reported in this chapter may seem reprehensible. Yet, as I suggested in earlier chapters, the unanticipated results that represent the animal's own inventive solutions and can often be reported only anecdotally are sometimes as important as the more traditional analyses of expected outcomes.

Enriched environments allow students and researchers to gather exciting data, much as they might in field work. Watching the ways in which complex animals solve tasks in noisy, less restricted milieus than a conditioning room or chamber does not produce the well-organized data that one is accustomed to seeing in behavioral journals. But it does provide an opportunity for observers to investigate ways in which animals derive unique solutions, rather than simply measuring their activities in a narrow response mode largely dictated by the experimenter.

Allowing animals to work together and set up their own mores and social structures provides observers rare opportunities to examine species and group differences in the establishment of social roles in captivity. There have been thousands of passive investigations that have looked at the behavior of captive animals in zoos and wildlife parks. Some of these have illustrated that species-typical behaviors do occur in a limited fashion using traditional zoo protocol.

Other investigations have emphasized the great incongruities between the behavior of animals in the wild versus those in captivity. For example, most of the literature on wild wolves suggests that within a pack, only a single female ordinarily delivers offspring in a given year, and that this is usually the dominant female. Yet, we have observed sibling matings, multiple litters, and cross-fostering within the same year in the Washington Park Zoo (Paquet, Markowitz, Sullivan and Bragdon, 1979). More comparisons between captive and wild behaviors of specific animals are essential in evaluating the appropriateness of the zoo environments. Our point in this book, and in this particular chapter, is that unless captive animals are given some special ways in which to manipulate and control their own environments to some extent, many of the comparative analyses become relatively meaningless.

The following chapter concerns apparatus design and selection and emphasizes the need for response flexibility. One measure of an animal's relative freedom is the extent to which it can choose the ways it wishes to obtain its reinforcers. It should be possible for thoughtful planners and investigators to provide increasing opportunities for animals to outwit us as they exercise some control of their own lives.

13
ABOUT APPARATUS

Many of the projects discussed throughout this book require apparatus that cannot be assembled from items readily available off the shelf. In those cases where commercial hardware might be appropriate, budget limitations often make it impossible to use prefabricated equipment. For those readers interested in conducting behavioral enrichment projects of their own, there are a lot of hard decisions to be made concerning the best way to proceed in material selection and construction. This chapter will deal in a general way with these problems and with some partial solutions which have been found useful in our work.

PRELIMINARY CONSIDERATIONS

Before one of our investigators begins apparatus development, we ask the following questions:

1. Should this researcher manufacture prototype apparatus?
2. Are the gadgets proposed potentially beneficial to the animals, research, and display?
3. Has sufficient time been given to identifying species-special requirements?
4. What special requirements does the physical environment demand of the equipment?

What follows is some of the rationale for these questions.

Should this researcher manufacture prototype apparatus?

In the manufacture of prototype apparatus, all of the traditional worst-case laws seem to apply—e.g.: if anything can go wrong it will, everything takes longer than it really does, etc. Consequently, it is very rarely that I have seen even the most modest-sounding timelines for completion of apparatus actually met. The builder should be prepared to expect delivery times occasionally to be very long for unusual components. Even with reliable suppliers, promised delivery dates are often exceeded.

Many researchers are willing to spend the time necessary to learn special techniques of soldering, welding, molding, etc., and for some rare individuals, all of these skills may be combined so that they can do much of their own construction. However, for the vast majority of workers, this is neither a cost-effective approach nor does it lead to the most reliable final products. In the area of electronic apparatus, for example, the day-to-day production of ever-more powerful and inexpensive equipment means that information only a few months old may be inadequate in the selection of the best components for each job. On the other hand, even if investigators do not produce their own equipment, this should *not* prevent them from acquiring a general working knowledge of each area of design so that intelligent questions may be asked of manufacturers or technicians employed in the work.

The major questions involve cost, availability, compatability with present equipment and personnel, repairability (use of standard non-esoteric parts), ruggedness (immunity to mechanical, electrical, and water abuse) and safety. If the apparatus is to be used with an existing system or if there are plans to expand the system later, it is important to select equipment with compatible specifications such as voltage requirements, logic levels, and plug and connector types. Interfacing widely differing equipment types may easily cost more than the equipment itself. When in-house construction is contemplated, all of the above considerations apply and other potential problems arise.

Calculating costs is a difficult problem. The cost of staff time for design and construction can easily exceed by many times the cost of parts for the project. Much too often, staff time requirements are neglected in the calculation of project expenses. The design of the equipment must guarantee that individuals other than the designer can understand its operation clearly and repair and maintain it in the field. It is imperative that a master set of schematics with trouble-shooting guides be available as soon as the apparatus goes into the field. Finally, of course, a well-equipped workshop with all required tools is essential. Running from place to place to borrow equipment for construction can be an absolutely debilitating procedure.

Many readers will undoubtedly see this section as a challenge, and, if they are like me, will go out ready to tilt windmills both literally and figuratively. If you are one of those individuals, be sure to work hard to find confederates who can actually accomplish the tasks that they promise. Several times during the last 10 years, people have asked to contribute to our work by designing or building specific components for projects which we had proposed. I do not remember a single instance in which those were completed as promised by the volunteers. They were almost always finished or totally redone by myself or our staff in time that could not comfortably be afforded. Find reliable coworkers and a means to pay them an appropriate salary. You'll be rewarded with much greater efficiency and the maintenance of your sanity.

Are the gadgets proposed potentially beneficial to the animals, research, and display?

Sometimes it is taken for granted that, since most captive animal facilities are relatively sterile, anything one can do will *de facto* be better than the *status quo.* Assuming that the equipment is safe for the residents and humans who come into contact with it, change for the sake of change may be good in a trivial sense, but surely, it is a terrible waste of time and potential resources compared with what can be done with a little more consideration and study. For example, I was once called by a friend who asked to send some of her students to train in our laboratory. She indicated that they already had designed and were prepared to begin installation of some new equipment and needed our help to determine how best to utilize it. Eventually the students did come to study with us and did some remarkably good work, but this was after I had successfully discouraged them from installing the equipment described in the telephone conversation. Their original plan was to put in a lever for an orangutan and a coin box for the public. The public was to deposit a quarter in return for which the orangutan could sit in a single location and press a lever that delivered food. This seemed almost certainly destined for failure. With the tremendous complexity and inquisitive nature apparent in the orangutan, it was highly unlikely that interest in such a trivial task would be maintained under the nondeprivation protocol that they planned. Further, if someone did want to use public donation-initiated opportunities for the animals, such an awesomely simple task was not likely to provide much income. Rather, this seemed like a perfect design to elicit visitors' complaints about being exploited. Even though my original condemnation of this project to my colleague was more volatile than it probably should have been, she took it positively and, along with her students, produced a number of significant contributions to captive animal enrichment over the years.

The importance of providing a fuller life for the animal should come first. But this does not preclude the careful identification of useful research questions that may be integrated with each enrichment project. For most of the exotic captive species, there is an almost infinite number of potential questions about factors such as sensory capability, circadian rhythms, cooperative versus individual responding, and ability for complex learning. With a little ingenuity, these can be easily dovetailed with new feeding or exercise opportunities for the residents.

In addressing the question of adequacy of display from biological, educational, and recreational standpoints, an important first step is to observe typical visitors' responses to animals in the current environment. It is well-established that most zoo visitors do not study traditional zoo graphics, and it is not appropriate to assume that they know even the brief information provided in that form. The behavioral enrichment designer should identify those things that are most importantly missing from the visitors' appreciation of the species' capability. Often, apparatus can then be designed to maximize the chance for the animals to exercise these natural gifts and to teach the public about their abilities.

Obviously, it is an ambitious undertaking to attempt to satisfy all enrichment, research, and display needs in individual projects. What can be more readily accomplished with adequate planning is simultaneous improvement in each of these areas. Visitors are remarkably entertained by the chance to observe experiments initially designed to explore the animals' complex learning interactions, for example the diana monkey exhibit described in Chapter 4. At the same time, they gain much greater respect for the animal's capabilities (Chasan, 1974), and the animals are provided entertainment and healthful exercise.

Has sufficient time been given to identifying species-typical requirements?

Besides the *general* importance of studying the species *prior* to planning behavioral enrichment, there are *special* reasons that pertain to apparatus. Whatever time is afforded to carefully survey the literature will be more than recompensed in preventing design blunders. Besides trying to develop apparatus which maximally displays behavior frequently seen in the wild, consideration of factors such as the animals' strength is imperative.

Some specific examples may help to make the point. There is some suggestion in the early literature on comparative learning that gibbons *(Hylobates lar)* may be less adept than other apes in learning discrimination problems. An acknowledged shortcoming of the physical testing situations was that these situations failed to take into account typical natural response modes and postures for gibbons. In both our sequential learning and discrimination testing situations, gibbons have proved to be exceptionally adept students. Beck (1967) has similarly found that moving a traditional patterned string test from a horizontal to a vertical plane was sufficient to greatly improve gibbons' performance. We believe that part of the gibbons' success with our apparatus was due to the installation that made it easy to respond with whichever limb they chose and from a variety of postures.

There are many tales about the strength of great apes that should encourage apparatus designers to build durable equipment. In Portland, we watched an orangutan *(Pongo pygmaeus)* bend a waterpipe away from the wall with one arm. In another zoo where they installed artificial turf, apes literally used it as a handle to lift portions of concrete off the floor. It should be clear that transportation of ordinary laboratory equipment to the zoo will not suffice for testing any of the most interesting inhabitants.

Wherever possible, equipment should be designed to increase the exercise and physical well-being of the animal. Sometimes providing an active chase or a challenging mobile task may carry with it potential danger of injury. A number of students and coworkers have proposed designs in which ground or near-ground level tracks would be used to run artificial animals within exhibits so that residents could enjoy chasing them. This sounds like a simple and useful design for many species. It has certainly been of utility in the special circumstances at dog tracks. However, placing such equipment in a home environment where animals have relatively constant access to it is especially problematical.

Electrically energized track would require constant maintenance because of problems with weather, dirt, urine, and feces. Mechanically pulled artificial animals that are safe are extremely difficult to design because they must be strong enough to work in spite of regular encounters with dirt on the running surfaces, and yet they must be able to be instantaneously stalled should they come into contact with a fragile part of the zoo animal such as paws, nails, claws, and tails. In designing the behavioral enrichment equipment for the tigers in the Panaewa Zoo in Hilo (Chapter 10), as much design time and equipment expense was spent in providing a solution to this simple-sounding problem as was required for construction of the entire computer system for the exhibit.

As suggested earlier, if new leaping, brachiating, or other gymnastic opportunities are provided, it is critically important to ensure that factors like landing surface receive proper attention. In three or four cases, initial designs brought to us would have required the animal to go through unusual contortions to avoid bouncing off the walls if they were to efficiently satisfy response requirements suggested by the experimenter. Even where simple improvements like ropes for primates to swing on are being planned, there are cautions to be observed. All possible excursions on artificial vines should be checked beforehand to make sure that the animal is unlikely to come into violent contact with hard or sharp surfaces. For some primate species, such as young chimpanzees, the use of very long ropes in unattended cages is hazardous because the animals may, in the course of play, literally strangle themselves. A simple partial solution to this is to secure both ends of the rope so that only safely limited twisting and excursion is possible.

Different designers and zoo administrators will inevitably come to different arbitrary decisions about the extent to which the animals' new behavioral opportunities should be tailored to maximize public education and recreation as opposed to being designed to increase freedom of choice for the animals. While in some cases there may be no conflict in providing ample opportunity to satisfy both these goals, for a larger number of species what would be nicest for the public may not be best for the residents. Many canine species, for example, require periods of privacy and opportunity for retreat from scrutiny in the face of perceived dangers for maximum well-being. Some hunting animals thrive on the opportunity to hide and pounce upon prey (or even upon dead food provided them). Providing these "retreat" opportunities for animals engenders realistic worries in zoo administrators about visitors' reactions to exhibits that often appear to be empty.

With a little imagination and luck, exhibits allowing privacy may actually bring surprisingly positive responses from the public. Obviously, dynamic, attractive, and interesting graphics that explain the animal's probable whereabouts are important. (Static, old-fashioned ones will not do as clearly shown in the famous unicorn experiment where an empty exhibit was clearly labeled "Unicorn—a mythical beast" and adult visitors continued to tell their youngsters that the animal was probably just asleep somewhere!) In the Sacramento Zoo, Bill Meeker designed and supervised the construction of a cheetah exhibit with plenty of hiding places. Although constructed on a modest budget, the exhibit is attractively naturalistic in appearance. Even though the cheetah are frequently difficult or impossible to find, young visitors delight in visually hunting for them.

With the use of one-way glass, lighting of a color relatively invisible to the animals, and other display techniques it is often possible to allow visitors an opportunity to see active species-typical behaviors and simultaneously give the animal a feeling of privacy. The Arizona Sonora Desert Museum includes a number of exhibits in which cross-sections into naturalistic domiciles are exciting to visitors while the animals largely ignore their presence.

What special requirement does the physical environment demand of the equipment?

Before design or construction of apparatus is begun, the area of installation should be surveyed with the help of maintenance and keeper staffs. This should include a careful review of the routine operations in the area. Most projects will require electrical power and careful waterproofing. In situations where easy exhibit entry is not possible, it is desirable to have the completed apparatus removable and accessible from outside the animal enclosure. Zookeepers and veterinary staff should be solicited for suggestions about the location of control equipment, wires, feeders, etc., so that cleaning, feeding, health service, and other maintenance operations are not hampered. Such consultation is also valuable in establishing a rapport with zoo staff.

In most cases, keepers can supply generous amounts of information on the daily activities and habits of the animals. One should not take as gospel that the old way is the only way, but thousands of hours of personal study can sometimes be saved by being a good student with a keeper as your mentor. They may save you from installing valuable equipment that the animals are almost certain to destroy because of their routine habits, or they may provide simple advice with which to guarantee the safety of your procedures.

Be sure to stock spare parts, and make yourself familiar with ways to easily unplug components for repair. Even when supposedly impossible (or "unconditionally guaranteed"), any part of the apparatus may break. Switches will get stuck, relays will wear out (it doesn't take the animals long to cycle electromechanical equipment through the millions of operations which comprise the useful life of these parts), transistors and lights will burn out, wires and connectors will break off, wires (for strange reasons) will lose the ability to conduct electricity, and animals more clever than people will find ways to remove explosionproof covers. Both 110 volt ac and low-voltage control wires must meet electical code standards that are often more rigid for public facilities than for homes or private laboratories.

Operanda, signal lights, and other equipment exposed to animals must withstand the forces that the animals will apply. Even small animals can exert tremendous pressures on common equipment. In glass-walled exhibits, any piece of apparatus torn away by the animal can become a dangerous projectile when hurled at the glass. Even more stringent requirements must be met by parts of the apparatus that interact with or are exposed to *Homo sapiens*. Lights and switches must be recessed so that they cannot be knocked off by a sharp blow. Housing should be constructed of 14-gauge or heavier steel and strengthened with angle iron inside or backed with quarter-inch steel plate. Such housings should be securely anchored so that they will not move when an adult jumps up and down on them. Unfortunately, what sounds terribly pessimistic here is factual. Virtually every place that exhibits wild animals has a majority of visitors who are kind, thoughtful, and appreciative of the unique opportunities that they're experiencing. But with even moderate-sized zoos having attendances of more than a half-million people per year, it takes only a small percentage of lunatics or genuinely malevolent individuals to destroy the products of hard work. By its very nature, behavioral enrichment equipment tends to be more complex and potentially fragile than most standard zoo furniture, so special protection must be provided.

CHOOSING THE TYPE OF APPARATUS

As indicated in the beginning of this chapter, the selection of electronic hardware has become a virtually day-to-day matter in terms of cost effectiveness. Any section dealing with specific components would unquestionably be outdated long before this volume goes to press. So, discussion of apparatus and selection of components will emphasize general *types* of equipment to be utilized. For ease of discussion, four general categories will be distinguished. These are: response registering and signaling, reinforcement delivery devices, control, and data acquisition.

Response Registering and Signaling

Response registering includes the levers, switches, photocells, contact detectors, strain gauges, etc., through which the animal communicates information to the apparatus. Most traditional equipment has used panels, levers, or other manipulanda with microswitches attached to register and define the responses. With work in the zoo, we have found a number of situations in which various types of capacitance detectors have been much more useful. For use in Portland, we designed a universal response panel with contact detectors that could be tuned to the animal's body capacitance. Thus, we were able to measure the discrimination capabilities of widely varying species, with essentially the same response effort requirements.

Integrated circuit advances have made the price of narrow band ultrasonic detectors (which do not require tuning for detection of proximity) competitive with capacitance switches. Today, these devices provide a more optimal solution to response registration without overloading research budgets. Obviously, caution has to be taken with respect to the range of the "ultrasonic" signal employed. Many of the most popular research animals have hearing that extends far beyond the typical 20-kilohertz upper limit in humans. Where these species are involved, radio-frequency field generators and receivers may provide a similar solution to problems of response detection, but at somewhat increased cost.

In the token economy described for diana monkeys, a light and photocell detector were used to count and record tokens dropped through the slot to order food. While this system worked excellently in the particular dimly lit, relatively dry environment where it was employed, we have found that a majority of zoo environments preclude the efficient use of photocell detectors. Any place where the animals, keepers, or weather factors are likely to deposit dirt will produce great difficulties. When a light is in the cage with the animal, it must be substantially shielded and not permitted to heat up to burn an unwary animal. Outdoor zoo animal environments have different light levels seasonally and with time of day. Photocell systems are seldom the method of choice in such field situations because of the amount of engineering necessary to develop well-focused, ambient light-shielded systems with apertures resistant to clogging by airborne dirt. We have even found that clever animals may learn to permanently block light sources if researchers aren't constantly vigilant!

There are a number of considerably more exotic methods of response registration that require more expensive specialized equipment. Three techniques that have special promise for zoo work are telemetry, radiological labeling, and weight detection. A significant problem in research dealing with multiple animal groups is the precise and automatic registration of responses by individuals. In those species where it is practical to put on collars or to place subcutaneous implants, an elegant and direct solution is to provide each animal with a signature transmitter that provides a signal clearly distinguishing it from all others in the environment. A tuned receiver can then be placed near each response station and data acquisition devices can then register both the respondent and the response. The cost of telemetry and its efficiency have undergone the same miraculous improvements as most electronics. Components can now be made to be virtually invisible. New systems that incorporate passive transducers, and integrated circuit controls and amplifiers which consume little power, reduce difficulties in design. However, battery life and size continue to be the limiting factors in virtually all field telemetry applications.

There are, of course, a number of fringe benefits to the development of a telemetry system because it can be expanded to monitor vital bodily signs and other internal parameters for purposes of health care and research.

Radiological labeling is another rather straightforward procedure that might be used depending upon local equipment available. This would involve the selection of long-lived, low-level isotopes that could once again be attached to the animal or subcutaneously implanted so that radioactivity detectors could register the proximity of individuals. Proper selection of components would make this a safe, durable system with few routine service requirements. Using radiological labeling in public facilities requires appropriate licensing *and* clear communication to visitors of the nonhazardous nature of the materials employed.

Perhaps the method of choice for those animals in which there are identifiable weight differences is the use of individual weights as signatures. Complete electronic scales, including digital readouts, are available for as little as $100. With careful selection, researchers can identify units that allow removal of the digital readout so that the readout may be installed at a distance from the weight-detection unit. Since the signal that activates the readout is electronic, it is relatively easy to transform this output into a direct computer entry.

One general installation that we are currently planning uses the digital scale readout in an area where both zoo visitors and researchers may observe the weight of the animal when it is on the load cell platform. All electrical wires and connections to the scale are to be made through a rigid conduit to eliminate dangers of damage to the apparatus or hardware disease for the animals. The electronic signals will also be fed to data-reduction devices so that there is a constant monitoring of animal presence on the platform. We look forward to many advantages with this system including the chance to observe changes in animals' weights as a function of age and nutrition, the accurate detection of individuals without the requirement of implants or attachments, and identification of sequential or simultaneous presence of specific animals at critical places in the cage where the weighing platforms are installed.

Although the input detectors for this system are no longer prohibitively expensive, maximum use of information from these inputs does require sophisticated digital equipment to analyze and predict slopes of weight change. There will also obviously be some occasions where, for finite periods, animals will have indistinguishable weights, as in the case where a juvenile in a dimorphic species may begin to overlap with the weight of the smaller of its parents.

Reinforcement Delivery Devices

Unlike the laboratory, where it is relatively easy to arrange the delivery of solid or liquid foods so that animals do not ordinarily have access to the equipment, the zoo sometimes presents formidable problems. In developing naturalistic feeding opportunities, it is often desirable to have the food appear away from keeper passageways and other areas where it might easily be protected. There is the further complication that, in many zoos, virtually every species has a somewhat different diet and requires special handling capabilities. Feeders must also present as few obstacles as possible to routine husbandry and health delivery needs, and this often makes it impossible to use commercially available feeders.

The delivery of liquid foods allows for a wide variety of installations using various configurations of tubing that may lead to virtually any area of the exhibit. By using appropriate combinations of gravity or pumps and solenoid valves, it is possible to accommodate liquids of varying viscosities.

As long as it remains relatively homogeneous, liquid food is easy to transport and to deliver through comparatively simple mechanisms. While these are outstanding advantages, there are some problems associated with liquid foods. First, there are some species for which liquids are simply not appropriate. Especially where long-distance delivery of food is involved, entire systems have to be flushed on a regular basis. Where feeding takes place throughout the day on an unpredictable basis, liquid subject to spoilage or contamination requires specially treated (e.g., refrigerated) tubing and terminal sections. We have found the greatest utility for liquid feeders in work with species that consume relatively thin solutions over reasonably short periods of time in areas that allow feeding above ground level.

Problems associated with the delivery of solid foods are primarily related to cost factors, space needs, and reliability. Pellet feeders so commonly used in the laboratory are impractical in zoos because they require specially formed, relatively expensive, shaped food. These dispensers are also notorious for requiring service at unpredictable intervals when food, broken by mechanical action, causes them to jam. In zoo work, I would recommend these dispensers only for those very special cases where constant attention to the apparatus is logical, such as short-term projects in which observers are able to attend to both the animals and the apparatus.

Universal feeders that use brushes or other motor or solenoid-driven arms to knock food from "lazy Susans" have proven exceptionally reliable, assuming that appropriately selected units are purchased. Manufacturers provide options for marine environments and for various-sized foods. Unfortunately, these reliable feeders are also typically quite bulky and require about one meter in each direction for installation. The architecture in most established zoos does *not* include keeper passageways or access rooms that allow convenient installation of this type of food-delivery mechanism.

Because of these considerations, we designed the simple conveyor belt system mentioned in Chapter 3. This easily manufactured feeder has been used in a wide variety of climatic and geographical areas without a single malfunction and has sometimes served for more than seven years. The design, in brief, involves dadoing a slot approximately four centimeters wide by seven millimeters deep in a piece of kiln-dried two-inch by three-inch lumber of sufficient length to support the desired quantity of food. An endless loop of 35-millimeter mylar film leader, approximately 0.3 meters longer than twice the board length, is constructed by lacing and hot-splicing the ends together. A 35-millimeter sprocket (available from commercial motion picture projector repair firms) is secured to the shaft of an appropriate motor. We have found the most reliable, least expensive motors for this purpose to be those manufactured for vending machines such as coffee dispensers. The motor-sprocket assembly is attached with a steel bracket to the side of the piece of lumber so that the top teeth on the sprocket are in line with the dadoed slot and about one centimeter from the end of the board (Figure 13-1). At the opposite end of the apparatus, a spring-loaded idler is used to maintain appropriate tension on the film belt. Depending on the length of the feeder (we have made some as long as eight meters), it may be necessary to provide rollers beneath the piece of lumber at appropriate intervals to prevent the belt from sagging excessively.

A final word about this simple construction: the majority of installations will be against a wall behind an exhibit or in some similar flat area, and it must be ensured that the brackets used for mounting do not preclude the removal of the film loop for sanitation or replacement. This means that *all* mounting hardware, including that for the motors, idlers, and the apparatus itself, should be attached to

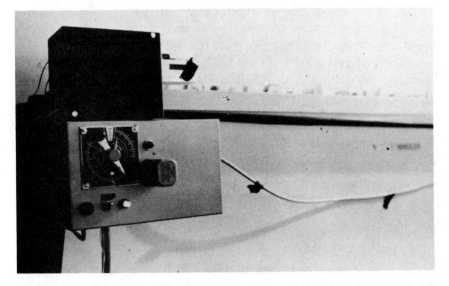

Figure 13-1. Typical feeder-belt installation in Portland Zoo. *(Photo by B. McCabe)*

the same side of the board. Since motors are readily available for clockwise or counterclockwise operation, a little advance planning will make this task easy. When properly installed, the belt can easily be removed (by depression of the idler spring) and washed in a solution of appropriate detergent.

Whether the belt feeder just described or some other solid feeder is used, it must be remembered that all animal facilities have visiting insects and rodents, and protective covers should be designed to minimize access to the food by those for which it is not intended. Where food is subject to short-term spoilage, we have found it convenient to mount feeder belts in chest-type freezers that can easily be converted to operate as refrigerators.

There are a number of types of reinforcement other than direct nourishment that can be used to encourage activity or provide new opportunities. Some that have been used occasionally in our work include programmed light, temperature cycles, or both, access to other animals for social activity or mating, motion pictures, toys, and opportunity to see, hear, or smell conspecifics. These alternative reinforcers are often more fun to work with than traditional food and water rewards, and they have the advantage that many animals will readily respond to novelty without the requirement for food deprivation. For some species that are occasional or unpredictable eaters, these methods also may provide more reliable performance. Especially interesting effects may be developed by illustrating thermoregulatory behaviors in cold-blooded animals. When integrated with appropriate graphics, such apparatus can provide a dynamic zoo education experience.

The use of punishment or negative reinforcers in general has been omitted from this section. This is because of my personal prejudice that there are few, if any, circumstances where automatic delivery of punishment is necessary or appropriate in the zoo.

Control

Of all the difficult apparatus decisions, the greatest ones probably reside in the areas of instrumentation control and data acquisition. Here it is especially easy to be penny-wise and pound foolish. With the new microprocessor technologies, it is plausible to dedicate entire computer systems to single projects. This helps to reduce one of the earlier difficult decisions: whether to bring individual projects by cable to a central computer, or to use less adequate local programming. The traditional concern has been that if a central computer required servicing, all local projects might simultaneously be disabled, but one hated to bypass the computer's power and use less effective individual hardware. Microprocessor chips can be purchased for less than $20 from some manufacturers, and complete systems that outperform computers costing hundreds of thousands of dollars 15 years ago are available for $5000. Except for temporary projects, it does not make sense to build your own system. Most companies have field consultants whose time is dedicated to consulting with researchers about the best configuration for their experimental requirements.

All this assumes that you have a budget, and unfortunately many potentially productive zoo researchers have to start with pocket money. Many of our students have found a "gold mine" in state and federal surplus agencies where thousands of dollars in overstock, "outmoded" equipment, or equipment from discontinued projects could be purchased for very few dollars. A valuable experience may come from inventing ways to interface esoteric equipment so that it does the required job. Where relatively infinite time and patience are available, scrounging may work effectively. Prototype equipment built on a budget typically suffers from idiosyncratic service needs (Figure 13-2). If there is a chance that the builder will leave, you may spend inordinate time trying to identify the source of malfunctions.

Whether user-built or commercial equipment is used for control of research parameters, the first major question will be the number of input/output ports to be provided. In general, it is a good rule to carefully identify all of the possible devices that might be interfaced with the equipment and then multiply that number by at least two. Once work has begun on a project new ideas will emerge, and it is much easier to connect through existing channels than to add in a patchwork fashion. Similar considerations apply to the translation and the storage sections of control equipment. Selecting a large enough component enclosure, so that additional circuits may be plugged in, is the minimum step required in order to obviate having the equip-

Figure 13-2. Prototype of controller for otter exhibit illustrates "idiosyncratic service needs" (see text) with medusa-like appearance. *(Photo by P. Foster-Turley)*

ment dictate the limitations of novelty that can be provided for the animals.

Although it may take slightly longer to construct, keeping the equipment as modular as possible so that components may be replaced by simple unplugging and insertion of replacements will help to prevent periods in which the equipment is unusable. Many commercial manufacturers of behavioral control equipment are increasingly moving in the direction of interchangeable parts, compatible voltage levels, and comparable space requirements. This allows the careful researcher to select the most reliable, inexpensive units that will accomplish the required work.

In circumstances where several projects are planned in a localized area, two kinds of equipment may provide the most flexible and cost-effective means of control. A general system including teletype, switch-selectable or plug-board inputs can be used effectively in trying new ideas. Where it has been possible to do initial testing with general systems in our work, we have almost always discovered ways to improve the design before committing ourselves to permanently wired equipment.

The second, more permanent kind of component will depend on budget factors. Computer control systems are initially the most expensive, but also the most powerful in terms of complexity of design. Minicomputer units can, if necessary, control several exhibits at once and also keep sophisticated individual records. A nice feature is that enrichment paradigms may be changed dramatically simply by changing the computer program. Because of the minute electrical energy levels involved in microcircuitry, proper shielding and interfacing may require expert help. In most cases, older technologies involving relays and discrete transistors do have the advantage of greater noise immunity.

Data Acquisition

In early phases of each project, a combination of human observers and simple event recorders may be the best alternative. After this initial research stage, it is most effective to utilize technologies that provide rapid data reduction and analysis. In the ideal case, the designer may be able to dedicate on-line computer equipment so that results are available almost instantaneously. Caution must be used to ensure that power failures or other malfunctions do not cause the loss of valuable data. Many new components offer fail-safe memory storage, even in the event of power outages. With other equipment, it may be necessary to provide backup battery systems or other methods to automatically provide instantaneous alternate power.

For behavioral engineers working on tighter budgets, very efficient systems may be employed that allow daily data treatment. For the Portland mandrill project described in Chapter 8, a portable cassette tape data-transport system was designed. The equipment at the exhibit stored the day's data internally in inexpensive microprocessor memory. These data were "dumped" onto magnetic tape by means of a modest cassette recorder which a technician brought to the apparatus daily. The tape was then returned to the research center where other computational format computers provided quick reduction and analysis.

FINAL REFLECTIONS

The inevitability that much of this chapter would be outdated before final publication was mentioned earlier. A few examples of this year's technological breakthroughs may provide a hint of future equipment trends. Several manufacturers are marketing prototypes of battery-operated equipment that may be carried into the field and are capable of recording a full day's data for very complex behaviors. This versatile equipment, which can also be modified for control as well as data-acquisition purposes, is in some cases advertised for as little as $600 including solid state digital readout. Instead of worrying about tape limitations, it is now possible to store all of the information electronically for later interpretation by a "mother" computer system. Many researchers will undoubtedly be able to do quite adequate work with small portions of time solicited from a central computer system.

As if this were not mind-boggling enough, the variety of input/ output modes is even more amazing. Almost all of the manufacturers provide options for direct attachment to every imaginable standard computer terminal, including telephone systems that allow data transmission from the nearest home, office, or phone booth.

One manufacturer has sent me a preliminary specification sheet for a voice input, to be marketed for approximately $1000, that will truly revolutionize data entry. This device eliminates the need for complex codifications of data written on sheets or punched into keyboards. The user simply speaks to it, identifies new terminology two or three times, and the machine learns the words—including the particular inflections of the researcher. The testing prototype has reliability exceeding 99% recognition. Imagine being able to record data while never momentarily stopping observation to write or look at keyboards! Imagine being able to invent new terminology or sort out the redundancy in your current usage without having to rewrite data sheets!

The newest electronic revolutions will surely convert more of us to the use of prefabricated equipment, which I have repeatedly emphasized has a number of advantages in conserving valuable time needed for field work, library study, and baseline data acquisition. But this chapter should not be closed without pointing out that for those researchers who continue to design their own equipment for the enjoyment and challenge of the enterprise, there are compensations. Staff will be more familiar with proper operation and maintenance of equipment carefully and thoughtfully assembled in-house. Construction of equipment for specific purposes allows the designer to meet the rigors of the zoo environment. Last, but certainly not least, in-house construction can sometimes be used as a controllable source of funds to help support graduate students and research assistants.

14

ANIMAL HEALTH
CONSIDERATIONS

Zoos have surprisingly widely varied veterinary staff arrangements that are not always clearly related to the size or solvency of the park. Some medium-sized zoos and wild animal parks only use consulting veterinarians; others have both a full-time veterinarian and a number of assistants. The ways in which assessments of behavioral enrichment effects on animal health are accomplished will be partly dependent on the veterinary staffing and routine. Properly planned and conducted behavioral research should have veterinary consultation from the outset as a major component.

In those zoos where the veterinarian is a regular employee, he or she is usually one of the key staff members included in the approval or disapproval of potential projects. Thus, it is important to understand veterinary concerns and responsibilities in order to best convey to health workers the value of new projects. We have been very fortunate in finding a number of zoo veterinarians willing to work in a totally collegial and supportive way. Michael Schmidt, who was our veterinarian in Portland, coauthored a number of articles describing animal health benefits of behavioral research, (e.g., Markowitz, Schmidt, and Moody, 1978; Markowitz, Schmidt, Nadal, and Squier, 1975; Schmidt and Markowitz, 1977). Mike also was generous with guest lectures to my university classes and helped with research interns. The following section is largely taken from one of Dr. Schmidt's lectures.

WHY CONSULT THE VETERINARIAN?

The zoo veterinarian has primary responsibility for maintaining the collection in the best possible state of health. Administering the latest and best techniques, medicines, and surgical care available in veterinary medicine to exotic animals is extremely difficult. As with all animals, the first problem compared with human medicine is that symptomatic report is not available. The veterinarian must depend upon identification of "signs" of illness. A "sign" is an objective, externally detectable, indication of an animal's health, represented in such typical measures as temperature, blood pressure, and behavioral depression.

It was Mike's hypothesis that many wild animals showed few early signs of illness because concealing transient disorders might have evolved to serve as a protective device. That is, an adaptation that prevented predators from detecting weakness until the last possible moment might, on the average, allow animals to recover from even fairly serious problems without attracting chase by enemies. At any rate, by the time animals are noticed to be ill and brought to the zoo veterinarian for treatment, they are often in precarious states of health.

The zoo veterinarian's most important problem probably involves restraint of exotic animals for health care delivery. In many cases, it is impossible to make adequate diagnosis or treatment until the animal is actually physically held. Unfortunately, most animals become highly fearful and excited in the face of immobilization and this, combined with drugs or physical restraint methods, can easily add to the animal's problem or, in the worst case, lead to its demise.

There are a number of apparent reasons why preventive medicine is extremely important in the zoo. First and most obvious, the early detection of illness provides a better chance to save the animal. Thus, good preventive programs include training in identification of subtle changes in behavior, food consumption, activity, fecal composition, etc., that may provide the first signs of illness. A second critical factor in preventive medicine is veterinary participation in the design of animal areas. Logically constructed facilities can greatly aid in early detection of illness. Occasionally, in striving for naturalistic appearances, zoo designers include so many obstacles to routine observation of animals that severe illnesses may go unnoticed.

It may be important at this point to consider the difficult compromises that must be made in presenting the most biologically sound representation of the animal's natural habitat while satisfying requirements for monitoring animal health. When the exhibit's residents are rare or endangered animals, this decision-making becomes especially problematical. On the one hand, we would like to utilize these precious resources to teach as much as possible about the species. But, simultaneously, there is an obligation to improve upon nature in protecting the animal from disease and accidents. Indeed, this whole dilemma should represent a major focus of behavioral engineers interested in exhibit design.

In other parts of the book I have emphasized that improper planning can predispose the animals to injury; e.g., jumps that require awkward landings on hard surfaces, sharp or protruding apparatus parts, and protocol that reduces the ability for restraint must all be avoided. Veterinarians are especially sensitive to the needs for sanitation, and I have seen Mike, as well as many others, involved in lengthy debates with architects and other exhibit planners about the need for cages that allow complete cleaning. Sometimes we have all finally laughingly agreed that the designer's dream may be one in which natural materials, including decomposed fecal matter, are present as they might be in the wild while the veterinarian's dream is of a tiled set of walls, floor, and ceiling that are automatically sprayed hundreds of times per day to eliminate all potentially deleterious microorganisms. It has sometimes surprised me that people of good will can finally come to mutually acceptable agreement when they are working from such diametrically opposed biases.

The veterinarian will want to participate in decisions about any additions, including behavioral enrichment ones, that may cause increased aggression. There may be circumstances in which increasing ritualistic aggression levels, which do not involve severe tissue damage, may help to enhance the fitness of animals and thus be defensible in zoos. The veterinarian will almost certainly require the final say in this judgment. New animals, pregnant females, older residents, and those suffering the pressures of recent introduction to a group are a concern of the health care crew that must be shared by anyone planning new behavioral requirements. For example, exceptional athletic feats, which significantly endanger the nursing of young, ought not to be required of recent mothers.

These are some, but certainly not all, of the reasons why the veterinarian must be consulted as early as possible in the conceptualization of new behavioral opportunities. Some suggestions for behavioral enrichment planners in considering veterinary aspects follow.

VETERINARY QUESTIONS FOR THE ENRICHMENT PLANNER

The first major question is whether the individuality of each animal has been considered. The first step in addressing this question is the establishment of a good "history" of the animals involved from all possible sources (keepers, curators, veterinarians, dietitians, institutional records, etc.). In planning for an existing group of animals, one would certainly, for example, want to know if the reproductive female is easily upset or has a history of killing newborn in the case of minor disruptions. Researchers new to zoos are likely to be surprised at the frequency of such behaviors in exotic animal populations.

Another major question concerns the completeness with which planners have thought of *all* the potential, sometimes subtle, effects of their equipment. The answer to this question may require several days of careful, systematic analyses of new physical structures and all of the ways in which animals may come into commerce with them. Occasionally, I have seen researchers accustomed to short-term laboratory experiments forget that apparatus that is perpetually left with the animals may overheat or present other potential hazards. Similarly, sometimes incomplete consideration has led to individuals forgetting that animals given the opportunity to feed themselves in new ways may actually eat much more than is good for them. And, as a final example, in some zoos I have seen apparatus installed that absolutely precludes any opportunity for cleaning bacteria, parasites, vermin, and other inevitable pests that may infest it.

Can the cage be entered without impediment in an emergency? If not, there may be devastating results for the reputation of the researcher whose equipment is in the way. In the most difficult cases, where only one entrance to the cage is available and the most appropriate apparatus needs to be attached to this door or window, a little extra effort will almost always produce ways to provide for instantaneous removal. In our work with diana monkeys where the station dispensing food in exchange for tokens was mounted on the entry door, we used a swing-away system so that the feeding device could be disconnected with one operation. Often the simple procedure of using flexible connections safely out of the reach of the animals will allow the accommodation of complex apparatus without impairing veterinarian and keeper entry.

If the proposed apparatus involves feeding the animals, can they select an imbalanced diet as a function of its use? While mixed fruits and vegetables are nutritious for and attractive to many animals, it is also important to see that they get sufficient chow or other hard substances designed to keep their teeth in shape. Where control of eating has been returned to the animals, we have frequently seen periods in which the strongest or most adept would tend to eat only the most preferred fruits or vegetables while throwing chow and other necessary items away or leaving them for other animals. While this can partially be remedied by shifting proportions of each kind of food delivered, it is important to be sure that careful attention is paid to which animal eats what. In one rather astonishing case, we introduced oranges as a variety food for a gibbon who had never been exposed to them before and, for the first few days, he preferred to eat nothing else. Since his cage mate consumed most of the other food very quickly, our first real clue came on the second morning with the tell-tale diarrhea.

Besides concern about the proportion of each kind of food consumed by individuals, there is the obvious question about dominant animals controlling the feeding or watering in a manner that causes nutritional deprivation for less aggressive cage mates. Unless specifically included in data-collection techniques, this can be a rather subtle effect and go unnoticed until there are serious consequences. Sometimes animals also become aggressive with the apparatus, especially if it requires considerable effort for them to feed. Special care may have to be given to ensure that chronic apparatus will not cause damage; for example, the wearing down of a bird's bill or a camel's incisors.

Here, finally, is a list of questions that might be asked about what might happen if the apparatus can in any way be dismantled by the residents. Can they:

- eat it and get hardware disease?
- use it as a weapon versus cage mates or visitors?
- be cut or lacerated by it?
- cause part of it to fall on or trap a cage mate?
- become shocked or electrocuted?
- use it to get out of the enclosure?

The veterinarian's concern that these questions be answered *before* enrichment projects are undertaken may be viewed by some researchers as a form of obstructionism. In truth, as Schmidt frequently suggested to my students, learning to anticipate and prevent potential problems earns researchers increased respect from all of the zoo's staff.

VETERINARY-RELATED RESPONSIBILITIES

The most general criterion for each project reduces simply to the fact that the good done for the animals should by far outweigh any potentially dangerous effects. Among the things that the designer will have to demonstrate is that there will not be significant impairment of ability to detect in the animal subtle changes that indicate possible illness. Fortunately, good designs can actually enhance this capability—as we will describe below. However, at the beginning of any new protocol, the regular veterinary and husbandry staffs will inevitably have to delegate to the research observer part of the responsibility for detecting early signs of illness. To clarify this point, over the years keepers and health workers develop a "feel" for the daily routine of animals within each cage. Observing that this routine is not being followed is one of the "signs" used to detect possible illness. Since new apparatus may markedly change the animal's space usage and activity levels, obliterating the old baseline behavior, several hours each day must be spent in careful observation of each individual to assure that they are getting along well and to help the veterinarian establish new typical behavioral expectations.

Nothing will make research more immediately disliked by veterinary staff than gratuitous offerings of health diagnoses. This does not mean that every effort should not be made to give objective reports of changes in the animal's behavior, activity, alertness, respiration, stool consistency, and so forth. It does mean that these reports should not be couched in quasimedical terms. Subjective opinions by researchers concerning the causes of medical problems seen in the animals will very easily alienate the veterinarian toward whom they are directed. The same is true of excessively anthropomorphic interpretations of apparent health problems.

Thus, the best approach to win the respect and cooperation of the veterinary staff is to involve them early and constantly in all health-related aspects of the initial project and in any subsequent proposed changes, to report in objective, scientific terms changes in the animal's behavior that may indicate health problems, and to constantly focus on ways in which behavioral enrichment can dovetail with general health care enrichment.

SPECIFIC EXAMPLES RELATING ENRICHMENT TO ANIMAL HEALTH

In a number of other chapters in this book, I have touched upon ways in which health diagnosis has been aided or health problems overcome as a function of the behavioral enrichment program in Portland. Chapter 6 elaborated on the design of apparatus for polar bears and described that the male attained the typical appropriate profile of a polar bear with a normal physiological reserve of fat. After several months of this work, the specific manner in which this came about included reduction in stereotypic pacing and increased healthful food-gathering activities. Obviously, as readers will remember from Chapter 6, the reduction of aggression was helpful to the maintenance of healthy bears.

The elephant memory work described in Chapter 9 is a particularly critical example of a health-related fringe benefit from active research. When the work began, our focus was on retention of a discrimination over an interval of many years. But, when the work was completed, the most important outcome was the detection of a visual deficiency in two of the long-term resident cows.

Specific programs to test sensory capabilities in zoo animals are now beginning in at least two university-based programs. Since for most of the species that reside in zoos there is little if anything in the way of normative data, this is challenging work that demands long-term studies for each new kind of animal to be assessed. A potential life's work for many investigators interested in helping and learning about zoo animals lies in this area.

The mandrill project discussed in Chapter 8 was selected for early development in our program because we hoped to radically change the cage dynamics. Animal health considerations, especially with respect to the male's treatment of the females and his aggressive behavior in the presence of newborns, were the major reason for this selection. A study specifically designed to investigate changes in space usage was accomplished both prior to and after the installation of the apparatus (Yanofsky and Markowitz, 1978), and results have illustrated great improvement in existing social dynamics.

THE SERVAL PROJECT

One project, not discussed in other sections of this text, was conducted with the serval *(Felis serval)*, a long-legged, medium to small-sized cat found in the open forest regions of west central Africa. The serval is renowned for its beautiful capture abilities. Often hunting in tall grass, these cats can literally flush game birds from the bush and catch them on the wing. It can also pounce on ground animals with tremendous speed, and certainly a large part of the felid's attractiveness resides in these behaviors. But—as with other examples which have been described in length—most zoo visitors rarely, if ever, have opportunity to gain appreciation for the servals' abilities. With the incentive to display capture behavior removed, the serval becomes an animal appreciated only for its svelte appearance, at best. We decided to develop a system that would provide the serval an opportunity to chase prey and the public a chance to view the active behavior.

As a first step an extensive baseline study was conducted. This work, which took more than a year (Schmuckal, 1974), focused on the three resident servals. The outcome can be summarized very simply: there were long periods of no activity whatsoever compared with what one would expect from the field literature; ethograms indicated that the most prominent active behavior was stereotypic pacing. This was not terribly surprising, since the home environment of these animals was a concrete and steel cage with a glass front and a small feeding slot through which food could be dropped by the keepers.

Following the baseline study, we explored in a preliminary way the effect of spacing the food on the interactions between the cats. Since there is some sexual dimorphism and occasional intraspecies aggression, we wanted to be certain that new protocol did not lead to hazardous behaviors. The procedure simply involved simultaneously dropping three pieces of food from overhead into three different areas of the exhibit. Results were startlingly systematic. There was no aggression with the exception of an occasional hissed threat, but there was an absolutely rigid hierarchy with respect to the order in which the food was consumed by the servals. The male always ate the first few pieces of meat, and then one particular female gained second access to the food, even though the two females were almost exactly the same size and age. Only after the first two animals were largely satiated was the third cat allowed access to the food.

After a lengthy period in which we continued to look for possible increases in aggressive behaviors which might result in tissue damage, we were satisfied that this was not a significant worry and progressed to the calibration for our final objective: the flying meatball. This "prey" had been decided upon because we wanted an active chase for the serval, and it was totally impractical for political and humane reasons to introduce live birds or small ground animals to the water-closet-like environment that was the servals' home. The literature on servals, including published pictures, suggested that the major distances in their leaps were horizontal rather than vertical, and we planned to descend our meatball (ZuPreem in a sausage casing) to a level where they might comfortably jump and capture it a few feet above the ground. Surprisingly, on the initial trial, the servals, exposed to their first animate "company" in years, leaped more than two body lengths vertically to catch the meat (Figure 14-1).

Excitement from this initial work helped to firm up the zoo administration's decision to move the servals to a more adequate outdoor environment in which we continued to engineer methods for introducing flying prey objects (Figures 14-2 and 14-3). Eventually a Teflon rod "birdie" was flown back and forth through the air many times a day over a total path of 13 meters. When one of the servals contacted this birdie, a piece of Carnifare (a prepared carnivore diet) was automatically delivered.

I would describe this apparatus as only partially successful. While the servals did actively use it, it was not uniformly pursued by all of the cats. Since servals are rather skimpy eaters, there was not the motivation to exercise at the greatly increased level that we have described for some of the other species. But, from the standpoint of

Figure 14-1. Serval *(Felis serval)* leaps to "greet" first animate object introduced into cage. *(Photo by B. McCabe)*

Figure 14-2. Flying meatball attracts serval's attention. *(Photo by B. McCabe)*

health diagnosis, the serval project did lead to an important finding (Schmidt and Markowitz, 1977). A sudden change in the male serval's behavior, in which he halted at the beginning of one of the prey-capturing sessions, led to a detailed veterinary examination. This examination uncovered the fact that the male serval had a chronic diaphragmatic hernia which had gone undetected through quarantine and through its previous sedentary existence in the zoo.

Figure 14-3. In final version of apparatus used in Portland, serval "captures" rod that delivers food. *(Photo by H. Markowitz)*

A NEGATIVE EXAMPLE WITH A HAPPY ENDING

At the outset of this chapter, stress was placed upon the importance of providing fail-safe protection for the animals with respect to any new apparatus. Unfortunately, one of our own projects can serve as an illustration of the potential for harm which results from incomplete safety precautions. In the installation of the discrimination apparatus for the orangutans (see Chapter 7), it was necessary to run flexible cables from the door in which the stimuli were inserted to the programming and feeding equipment. Despite my repeated instructions that this cabling had to be protected with rigid baffles so that it was impossible for the orangs to reach it, the work was started without this protection in place. (This was, incidentally, the only case in which I can remember that a personal schedule conflict made it impossible for me to see new devices before we exposed them to animal use.) Harry, the male orang, managed to snatch a piece of the cable in the interval between one of the testing sessions and sat looking proudly as he chewed upon its outside cover.

The veterinary staff and I were immediately contacted by an alert keeper, and we raced down to see what remedial action could be taken. Harry and Inji were isolated in a small holding area adjacent to the exhibit, and I proceeded to try to barter with Harry for his new toy. But, as was emphasized in detail in Chapter 12, other animals frequently outsmart *Homo sapiens.* I had no intention of hand- or armwrestling with Harry who, despite his gentleness, could easily have fractured me. We went through a bizarre hour in which the male orang was able to obtain a number of "goodies" by pretending that he was willing to exchange his piece of equipment for them and then, at the last minute, snatching back the cable as well. Through all of this, Inji sat pleasantly next to Harry, occasionally reaching to ask for one of the treats for herself and looking at me with what clearly appeared to be sympathy for my repeated failures.

Finally, in what would have been the most humiliating stroke of all had I not thought of these orangs as friends, it was Inji who saved the day. It occurred to me that she really wanted to help because she occasionally prodded Harry and looked in my direction. I finally resorted to asking her in a combination of plain English and gestures if she would get the wire from Harry. After I gave Inji a piece of fruit, she put her arm around the male orang, gave him the fruit, gently took the wire, and handed it to me through the bars. We all collectively breathed a sigh of relief.

While this happy and perhaps amusing ending is fun to relate, it might have ended tragically. Hardware disease is a terribly common problem in animal facilities, and occasionally leads to death. When electricians, cage builders, and other routine construction crews work in the zoo, we repeatedly implore them to be careful with their materials. Hediger (1950) placed a blow-up of x-rays showing the results of hardware disease—an animal's destroyed alimentary tract—into the workmen's area of the Zurich Zoo. Certainly those of us interested in improving opportunities and health care for zoo animals do not want to be responsible in any way for producing potentially lethal environments.

CONCLUSION

The surface has been barely scratched with respect to the ways in which behavioral enrichment designers can be of aid in veterinary matters. Carefully considered designs can help with *restraint* problems for virtually every species, while making the environment as comfortable as possible for the animals. Programs ranging from *routine husbandry,* such as elephant toe trimming, to *screening for sensory anomalies* may be enhanced by means of contemporary behavioral technology.

Hours spent brainstorming with veterinary and husbandry staffs about methods for improving health care are repaid in manifold measure: (1) enrichment designs can from the outset include equipment (which may be surprisingly simple and inexpensive) which will aid rather than impair veterinary procedures, and (2) after careful consideration, initially innocent-sounding devices may be discovered to be potential hazards for zoo animals. I believe that the future will increasingly see behavioral researchers devoting themselves to innovative ways to help with the assessment and maintenance of animal health.

15

SOME RULES FOR
RESEARCHERS

For a number of years, I was in the somewhat frenzied positon of constantly "changing hats" as I served simultaneously in the roles of director of the zoological center, associate director of the zoo, and university professor. Hectic as it was, this routine did serve to provide perspective on the feelings about research in several quarters. In this chapter, primary stress will be placed upon ways in which to approach zoo administrations with requests, the role of visiting researchers in animal parks, and some questions about mutual ethical responsibilities.

The premise in all that shall follow is that zoos should focus their research efforts almost exclusively on work that minimizes danger to the animals and is ultimately conducted for the animals' benefit. There are very few well-defined reasons for using exotic animals, such as those exhibited in zoos, in acute experiments or in ones that do irreversible tissue damage. In those unusual cases where medical models may make good use of animals, such as giraffes in high blood pressure studies or mule deer in studying red cell sickling, provisions should certainly be made to conduct studies with irreversible effects in specially prepared nonexhibit facilities. The question about whether these need be attached to zoos at all would require discussion beyond the scope of this book. It has been my experience that zoo-trained veterinarians are willing to provide extensive advice and assistance to medical research facilities that have well-defined reasons for the utilization of exotic species.

After initial visits to the zoo, the potential researcher might greatly benefit from taking the following steps prior to asking for an appointment with an appropriate administrator. First, a very clear conception of the areas of research for which training and experience have prepared the proposed investigators should be available in a brief resume that can be left at the time that an appointment is scheduled. Second, it is worthwhile to find the exact names (including pronunciations) and titles of those with whom one is likely to meet. This should also include a clear understanding of the responsibilities of each individual within the zoo staff. Third, it is very helpful to know about the exact fiscal and governing basis of the park. Although there is only one national zoo in this country, other levels of governmental ties range from state to county to district to municipal organizations. A number of zoos are administered by independent nonprofit zoological societies, and quite a few have combinations of both governmental and societal leadership. The particular form of administration may play a very significant part in the amount of red tape that has to be cut before research can begin. Fourth, as much as possible should be learned about the history of research within the particular facility in order to intelligently put new proposals into perspective.

There are two organizations that new zoo and aquarium researchers should seriously consider joining. The first is the American Association of Zoological Parks and Aquariums (AAZPA) that will be happy to send information about its goals and requirements for associate membership. (Send requests for information to: Oglebay Park, Wheeling, West Virginia 26003.) The monthly newsletter from this organization includes many facts of general interest, including information about new legislation affecting animal regulations and permit requirements, new exhibits and construction, special educational, research, and conservation programs, meetings of zoological interest, personnel changes in zoos, and a brief literature review. Attendance at the regional workshops or annual meetings of the AAZPA is an interesting, informal way to learn about and contact staff from many zoos. These meetings also typically have significant segments specifically devoted to new research ideas and findings. Proceedings of all of the meetings for several years are available in published form, and these publications, along with the *International Zoo Yearbook,* include materials that are not always referenced in the standard scientific abstracts.

The second organization that should be considered is the American Association of Zoo Keepers (635 Gage Boulevard, Topeka, Kansas 66606). The *Forum,* published on a regular basis, includes useful information about animal husbandry and general zoo innovations of special interest to those who come in direct contact with the animals. In a few zoos, there are open meetings of AAZK where keepers may welcome sincere visitors who wish to constructively participate.

It should be clear from some of the wording above that, like all complex organizations, each zoo has its individual character. Consequently there is no universal prescription for the best way to introduce oneself. However, in the great majority of cases for larger zoos, the first contact may be with a person titled curator, head keeper, assistant director, or zoologist. Despite the fact that I have emphasized the importance of having well-documented skills and ideas, it is also vitally important to enter the initial meetings with an open mind. If you can encourage the staff member to do a significant part of the talking, it is possible to discover wonderful avenues of research that had not been envisioned prior to the meeting. Additionally, it is of some importance to learn the right jargon. Selecting goals that are consonant with the stated goals of the staff will also greatly enhance the chances of conducting research.

Last, and perhaps most important, is the question of how to be funded. I have repeatedly pointed out that most zoos are always in some financial jeopardy, and whatever resources the researcher can bring to them rather than asking of them will be much appreciated. Sometimes this will simply amount to the fact that university or other outside equipment may be utilized in the zoo to benefit the animals or to help with data collection. Most often, significant programs will also bring positive media coverage and other beneficial public relations that enhance the zoo's image. It is important in such media coverage to remember to credit the institution for the opportunities given to you and to explain the nature of the work. Of course, in the ideal case where one can bring substantial funding to the zoo for the conduct of research, proposals are made much more palatable.

The researcher should be absolutely sure that any authorization given him has passed through all of the necessary channels in the zoo administration. Although sometimes this is cumbersome, it will preclude some potential heartbreaks and false starts, a point to which we will return in the section on mutual ethical responsibilities.

THE ROLE OF VISITING RESEARCHERS

If a person intentionally tried to design an environment with high probability to drive laboratory scientists mad, it might well resemble the typical zoo. In various parts of this book, I have mentioned keepers who inadvertently or intentionally dumped food into environments where animals were being tested on food-related tasks. This is only the tip of the iceberg. In the midst of critical data collection as you wait to see an unusual behavior expressed by an animal, you may hear that it is time for a break, to wash up, to go home, or to take care of any of a number of other "inalienable rights." In many zoos, if one wants to remain welcome, there is little recourse but to carefully plan research schedules to avoid the most remote chance that they may conflict with staff priorities or routines. This will be especially apparent when you first begin, because zoos maintain, more than most institutions, a sort of apprenticeship-by-fire in which newcomers (who may require substantial chunks of staff time or attention) are prime targets.

In spite of the negative tenor of the previous paragraph, I sincerely think it worth the investment of time to get to know zoo people at all levels. Once a worker progresses beyond this entry maze level, he or she will unquestionably identify keepers and other zoo workers who are extraordinarily helpful. Once active research begins, it will become clear that keepers, handlers, and trainers are the backbone of animal parks. They deserve your respect, and they can absolutely enhance or destroy active research projects. In a number of zoos, we have had people come in on their days off and even in the middle of the night to help with research programs that they thought were valuable.

The exact role a researcher plays will necessarily represent some compromise between the ideal goals that they envision and what is allowed them by the zoo administration and staff. But within this range of compromise, researchers should avoid bending excessively to tradition. In particular, I do not think that it makes sense to abide the notion that research must never interfere with current routine.

Researchers must involve themselves in *actively changing zoos* if they are to benefit the animals that reside within.

It cannot be too heavily emphasized that zoo personnel and governing bodies will appreciate courtesies extended by participating researchers. I have occasionally seen people who asked for and used research opportunities in the zoo and presented vast collections of superficial criticisms without the vaguest acknowledgement that the institution they were criticizing had been open enough to allow their participation. All too frequently, researchers may be discourteous in an entirely different sense, perhaps unintentionally, but it is nevertheless devastating to their reputation with the zoo staff: they may forget to share what they learned with the institution that provided the chance for study.

Much of what I have just said may have an excessively pedagogical tone, but I do not mean to preach to or to frighten potential zoo investigators. Instead, it is my hope that if you are one of those who wishes to begin research in the zoo, you may save valuable time by avoiding some of the potential pitfalls. One direct method for minimizing jeopardy is to commit as much of your agreement with the park as possible to written form—a point which will begin our next section.

MUTUAL ETHICAL RESPONSIBILITIES

After several years of admonishing visiting researchers to *complete* work they had initiated with enthusiasm, I settled on a formal plan which would have saved much anguish from the start. Even though we had always required a written proposal and the usual rough calendar and budget summaries, no explicit description of mutual responsibilities was included. I honestly thought at first that such contracts might be excessively formal and be seen as barriers against research.

Just the opposite is true. A well-devised set of statements identifying exactly what ground may lead to the termination of projects is essential for the protection of both the zoo and the researcher. Perhaps a few examples will help to clarify this need. The majority of new researchers come from training backgrounds where special facilities protect the investigator from most unexpected encounters with dangerous animals. Existing zoos are simply not constructed upon these same principles. Even long-experienced zoo personnel are occasionally injured when they walk through a passage corridor too close to the trunk of an elephant, the paw of a polar bear, or the enclosure of a venomous reptile. It is no wonder that there is great resentment and concern among staff when research workers wander into unauthorized areas. Many zoos cannot afford the multiple rekeying necessary to ensure limited access by researchers. So, if the research protocol requires that keys be given to experimenters, a contract must clearly tell them which areas are authorized to be entered.

Another critical element in formal research contracts involves authorized interaction with staff members. Some zoos have rigid protocol about assignments for staff at various levels. The contract should serve to inform the researchers about whom to contact for needs that involve staff time. In many zoos where this is not specified, morale may be seriously affected when senior personnel admonish keepers for spending time on research and not completing other expected duties. The contract will also serve to protect the researcher from occasionally being used as an excuse to avoid less desirable routine tasks.

Sometimes researchers, students, and volunteer assistants who help with the work may assume that they have less responsibility for punctuality because they are not employed by the zoo. Failure to keep appointments, especially those where staff have made special arrangements such as moving animals or clearing facilities for research, is a sure way to give experimenters a bad name in the zoo. The written contract should make clear from the outset that such behaviors are sufficient cause to cancel projects.

Clearly, there is protection afforded to both parties. Zoos are actively lobbied against by some vocal opponents, including occasional powerful members of the media. Although some of these zoo opponents exercise care and present well-reasoned viewpoints, many are anxious to find any source of criticism that may damage the zoo's reputation. Researchers, frustrated by being unable to complete projects in which they have invested time and effort, may be tempted to provide encouragement to anti-zoo elements. Consequently, if research is involuntarily discontinued, the reasons that led to this decision should be clearly understood. The contracts signed by both parties provide a public record and help to avoid many conjectural debates. Conversely, if bona fide research is being carried out in a conscientious and ethical manner, the investigator should be protected from tyrannical or whimsical elimination of programs once they have been agreed upon in written form. If specimens or historical data belonging to the zoo are used by researchers, it should be very clearly established where ownership, copyrights, etc., are to reside.

Finally, careful attention needs to be paid to those zoo decisions that may temporarily bring research to a standstill. A good example is the question about animal health. In our contracts, we arbitrarily made it clear that the veterinary staff always had final say about animal participation in any research activity in case of illness or other discomfort. If research protocol absolutely requires the participation of *every* animal *every day,* this last contractual provision needs especially to be brought to the attention of the investigator, because it would require a lot of time for the veterinary staff to examine these animals to see if they are fit each day; thus, the project may be unworkable from this standpoint alone. Some projects are simply not practical to conduct in zoos.

CONCLUSION

Zoos are unique and still barely tapped resources for research. In this chapter, I have tried to share some ideas about how to get acquainted, how to get along in a responsible way while researching in the zoo, and how to gain some protection from unethical treatment.

With the emerging trend to accomplish research with directly observable, practical value, university researchers may look to ways in which their work will be of utility to the zoo. Almost every zoo would appreciate well-edited student contributions to their public education programs. This might include materials for teaching machines, such as we have touched on in this book. Most certainly, researchers should include graphics (either temporary or permanent) in their fiscal planning to describe the nature of work which the public has the opportunity to observe.

Most zoos have large numbers of volunteers, anxious to work with the resident animals. Well designed research protocol, careful screening procedures to ascertain sincerity and talents of the volunteers, and detailed supervision are necessary to useful incorporation of volunteers in research projects. Good programs involving volunteers are welcomed by zoo administrators who are anxious to have encouragement and specific tasks provided for people who want to help the zoo.

Above all, if you are a person dedicated to improving the living conditions and appreciation of sensitive and diverse creatures, the zoo may be the place for you to conduct innovative work. The endless struggles against institutional inertia will be paid for many times over each time you see some animal measurably benefiting from your work.

16
THE FUTURE

The basic premise of what I chose to label behavioral engineering in the zoo nine years ago seems to be entirely compatible with what most zoos suggest is acceptable research and captive habitat improvement. Yet, perhaps partly because of the very name "behavioral engineering," there has been some objection to the work by zoo traditionalists. As we suggested in the early chapters, much of this opposition is clearly a matter of confounding tradition and nature. Interestingly, we have occasionally found that projects that people did not like under the title "behavioral engineering," are entirely palatable to the same zoo group when they were resubmitted under the name "behavioral enrichment" or "increasing opportunities for zoo animals." Whatever its final name, behavioral enrichment is still in its infancy in the zoo world, and there are many reasons to be hopeful about the future.

This last chapter will briefly summarize three ways in which I think the planning of responsive environments may best progress in the immediate future. First, there are the significant revolutions in instrumentation that no behavioral researchers can afford to ignore if they are to be sensitive to the importance of sharing data and if their projects are to be cost effective. Second, there must be a direct exchange of ideas with exhibit designers at the outset of planning. Not least important is the need for researchers to carefully attend to those species where the problems of captivity are real rather than imagined, a point which will be considered in detail below.

REVOLUTIONS IN INSTRUMENTATION

The advent of single-chip microprocessor units has changed the electronic designers' notions in about the same magnitude as the transistor and integrated circuit "revolutions" (see Chapter 13). Cost factors no longer preclude the use of minicomputers for each exhibit. Space considerations, which often led to the use of less complete and adequate instrumentation rather than to computers, have also shifted remarkably. In the space that one might have placed five or six relays and a few transistor drivers, an entire microprocessor assembly can be accommodated.

The most obvious advantage of using programmable equipment is that behavioral paradigms may be changed with a minimum of effort. With erasable read-only memories, a technician can remove the memory, place it under an ultraviolet source for erasure, and then write in a new program. For example, if an environment were designed for primate species, and this included a whole array of available activities and potential ways to have the animals "order" food, drink, or other reinforcers, one could easily add new equipment and new contingencies to the home environment without scrapping the control equipment. All that would be necessary would be to rewrite the program and to incorporate the additions.

Data treatment is unquestionably the most boring of the chores that confront most behavioral researchers. It is comparatively easy to stimulate students and fellow researchers to get out and directly observe behavior. It is somewhat more difficult to get investigators to carefully and systematically write down every critical event described in the data-collection technique. But what is nearly impossible is to train people to love the arduous task of compiling the results for statistical analysis. Much of this last work is unnecessary today because of the feasibility of directly storing data in a format appropriate for computer analysis.

One system that we have found very flexible is the use of high-grade magnetic tape cassette components. I have described this earlier for the mandrills' speed game, where one of our technicians daily "dumped" all of the data which had been stored by the microprocessor controls onto cassette cartridges. These cartridges were then brought directly to the Research Center where the computer analyzed their contents and printed them out for our permanent record files.

Although there are very apparent advantages for individual projects, the greatest bonus of using microprocessors and other integrated circuit equipment may come with multiple projects. It is possible to minimize the hardware at each behavioral station and to accomplish the majority of programming and data treatment in a central location. This may be done either by hand-carrying tapes back and forth as described above, or by direct telephone or cable lines between stations and the control center and even to the researcher's home.

The reader who has not thought extensively about electronic control or data processing may need some further amplification to prevent confusion between these and other computer models with which they are familiar. Many investigators are disappointed when they install entire laboratories controlled by single computers. This disappointment comes when they discover that with a single "brain" malfunction *all* of their experiments simultaneously cease. While it is important to have everything come to a central processor, it is demoralizing to be unable to turn to one experiment while another is "down." The spectacular advantage of the new electronic hardware is that it is now entirely feasible, within limited budgets, to have each station functionally independent and yet quite flexible. Should a researcher's major computer facility be out of order or closed for the weekend, the individual experiments do not cease but, instead, may continue to store their data until the master computer is ready for further communication.

In Chapter 10, while describing the tiger's activities in a general fashion, one behavioral paradigm based upon microprocessor circuitry was elaborated. Ron Fial, my good friend and electronics collaborator on many of our projects over recent years, designed most of the tiger project's circuitry. We have often spoken about one of the major design problems that may be of some general interest: it seemed that we always built prototypes! Each time that we finished a project, we thought of ways in which new hardware could be used to simplify or improve it. When we designed a speed game for the children's zoo in Hilo, we did not use the same circuitry as had been used in Portland. When we decided to build a new apparatus for some species for discrimination learning, although the behavioral paradigm was the same, we always changed the parts used. On the one hand, this seemed meritorious to us because we wanted to be cost effective and to profit by previous experience. But, in retrospect, it was difficult to break the habit of always building things in a brand new fashion. At about the time our work in Portland came to an end, we had finally begun to standardize with similar microprocessor and IC hardware, so that the new projects would allow direct data analysis and easy reprogramming.

There is nothing more humbling than watching the computer's efficiency at treating your data. Materials that took endless nights and days with hand calculators or computer cards or keyboards now flow from a high-speed printer more rapidly than you can read. Unquestionably, we would all be better advised to teach our students how to understand and work with computer methods for translating their data than to chastise them for not being willing to sit at a calculator for months.

At the very least, anyone planning behavioral enrichment projects will want to consult with an engineer or technician proficient with the latest hardware and design. Much money and time can be saved and research proposals can be much more effectively written in this fashion. Ideally, it would be nice for each individual investigator to gain a general familiarity with circuitry by looking at any of the new primers on integrated circuitry and microprocessors. There are already many citizens with terminals in their own homes, and it seems to us only a matter of time before anxious investigators may sit at their own home consoles and look at the latest results from a remote laboratory pouring forth from a video screen. There is every reason for zoo workers wishing to do the most for animals in their care to tune in the same channels!

WORKING WITH EXHIBIT DESIGNERS

There are many different types of exhibit designers that provide services for zoos. These range from standard architectural firms (including landscaping architects) through specialists in rock work and plastic facades. It is possible, by talking to experienced zoo people, to find a handful of individuals with excellent reputations for each kind of exhibit. But, there is also something to be said for starting with someone with fresh, new ideas. In the worst case, using the same

firms over and over may lead to all zoos looking alike. It has already
led to the development of some medieval zoos in the twentieth century.

It was really exciting working with zoo planners in Hilo because of
their willingness to be flexible. Jim Juvik, who directed the development, was anxious to find ways to increase the stimulation which the
captive environment would inevitably decrease. The architects, electricians, and city and county personnel were willing to help even
though they had never seen zoo innovations like these before. Undoubtedly, for the most part, we were just lucky to have had the opportunity to work with such fine people, but I would offer a few suggestions for ways to approach work with designers.

First, it is imperative that you share in *understandable* ways the
motives for your work. For us, that meant stating as clearly as possible that our designs were to be noncompulsive, i.e., that we wanted
to give the animals chances to do things when *they* wished and to increase the variety in their lives. One should not expect that designers
will have time to read endless reprints that you might send them or
lengthy proposals written in behavioral jargon. Instead, address the
problems and your proposed solutions in every-day language. Then,
be open-minded about allowing the designers to tell you if they think
"you're nuts!" Sometimes this is the best way to find out that the
designer doesn't really have any comprehension of your plans. Until
you progress to the point where you have some mutual understanding of the validity of the work, there is little hope that you will accomplish the maximum job on behalf of the animals.

Just as with most conventional kinds of building, things may change
dramatically as they go from planner to architect to builder. In the
worst case, this may be catastrophic. In one zoo where a new orangutan exhibit was accomplished, there was very little communication
between the animal specialists who had the original conceptions and
the final contractor. The contractor called the architect one day to
describe what he saw as an ignorant set of planning. Why place the
duct work up near the top of the moat when it would be much cheap-
er to install it halfway up the wall? The architect agreed to the seemingly perfunctory change without consulting the zoo administration.
The orangutan apparently approved the change, too, since it stepped
into the duct and climbed directly out of the exhibit. While this is

certainly an extreme case, it is not exceptional to have well-meaning people add much time and trouble by applying conventional wisdom to the unconventional problems of maintaining zoo animals.

Environmental engineers have as one of their major responsibilities the safety of the new devices to be installed. If some artificial hill is to provide an area for a rodent to burrow after prey, it cannot have sharp protrusions or exposed moving parts likely to catch paws. In most cases, it will not be sufficient to draw the plan and write notes describing these requirements. The designer will have to watch the construction of the hill throughout its fabrication. Otherwise there is an excellent chance that a significant percentage of prototypes will have to be scrapped or that compromises with animal safety will have to be made.

Much of this seems terribly negative, but it does have a bonus feature. Fledgling researchers in behavioral enrichment can learn tremendous amounts by hovering over their projects in unobtrusive ways. Seeing actual fabrication techniques, talking to architects and workmen about simpler ways to accomplish goals, and reviewing code requirements with inspectors is a real education. The open-minded planner may find much more efficient and attractive ways to accomplish enrichment projects by becoming familiar with standard construction techniques.

In this section, I have emphasized the need to maintain originality while learning about typical design and construction methods. One final important point about this intensive collaboration is that it may be the only way to guarantee that more than lip service will be given to the fulfillment of the behavioral planner's conceptions. Everyone who has worked on any new complex project knows that part of the success or failure lies with the cooperation of the major workers and supervisors involved. The first few days are spent in getting to know one another, understanding the idiosyncracies of each other's language, and assessing the real capabilities of construction crews. It has most often been my experience that the more projects accomplished with the same people, the better and more professional the products. Part of this may be because the specifier of behavioral equipment learns to be more efficient in design. But equally important is the increased convergence of ideas of the exhibit designers and craftsmen with the behavioral consultants.

SOME FINAL GENERAL CONSIDERATIONS

It is my firm feeling that behavioral enrichment should not be "over-sold." There are a number of species for which ample space and naturally available foods may be the best solution to captive environ-ments. It would be foolish to put something into every animal's en-vironment just for the sake of saying that you have done it. In this same vein, it is important to study carefully the purported problems in need of the behavioral engineer's attention. In their enthusiasm to be helpful, people often identify behaviors as aberrant when they are instead quite typical for the species.

For example, while there is much need to improve the size and to-pography of a majority of the captive wolf environments, the most commonly reported "signs of boredom" are a source of some amuse-ment for people who work with wolves. Every year, I see a number of letters to the editor describing the fact that wolves are "obviously bored and neurotic" because zoogoers have seen them pacing back and forth in their cage. A little attention to the field literature on this species would illustrate that pacing is a significant part of the wolf's life and that their tendency to stereotypically use the same pathway is so pronounced that they actually wear down wolf trails in the wild. It would obviously be possible to design programs to re-duce this pacing in captivity, but this would certainly not make their behaviors more natural.

Another common complaint that zoos receive is about the lions who are "always asleep." I have heard some suggestions from col-leagues about methods to have lions work much of the day. If the function of zoos is to provide education about species-typical behav-iors, then one should probably not initiate any extensive hour-to-hour requirements for the lion. I am afraid that what most people honestly know about lions has come from the circus and from exag-gerated movies and television descriptions of their capture and feed-ing behaviors. In the wild, they are much as they are in captivity, sleeping and lolling around a great proportion of the day. A possible solution to incorporating behavioral enrichment for lions in tradi-tional zoos might be to schedule and inform the public of the few times a day when some prey might become available, and then to simply allow the lions to capture the prey and loll around the rest of the time.

The limits by which species' environments may be enriched are defined only by the imagination of the researcher. But, there are some species that are "naturals" for immediate attention. Arboreal primates, ground-burrowing animals, and marine mammals are good examples of prime target groups for which to begin programs. For the most part, zoo exhibits do little to encourage use of natural capabilities by these species. There may be bars or artificial trees for the primates to swing on, but little reason for them to use them. Ground animals may show attempts at burrowing activities, but these would be greatly enhanced with the reintroduction of some contingencies similar to the wild requirements of food-gathering or predator-prey behavior. Most marine mammals will become active with the introduction of even the simplest devices or extra attention by keepers. Here the major remaining challenge is to increasingly produce naturalistic contingency repertoires and behavioral opportunities.

If there is one major point which I have tried to make throughout this text, it is my hope that planners will devote themselves to making environments that allow as much natural flexibility as possible for the captive species. There is a terrible tendency for some workers to want to press each animal to show just how much they can accomplish without regard to the comfort or well-being of the animal. I personally deplore exhibits that teach chickens to play baseball on command throughout a working day. Similarly, nothing makes me more uncomfortable than watching overfed spider monkeys sitting on the ground and begging for handouts from the public that throws junk food at them.

We can surely do better to provide naturalistic education and recreation. It is not necessary to deprive most animals to have them become active. Given the opportunity, most complex animals are anxious to work for the sheer fun of it. Administrators need not be so fearful of telling the public that it is not their "right" to poison the animals with garbage food. Those zoos and wildlife parks that have restricted or eliminated public feeding are among the best respected in the world. The prescription I offer future workers in behavioral enrichment is simply this: leave as many decisions as possible to the animals while providing them increased behavioral opportunities. Animals in nature are behaviorally beautiful, partly because of the variety of methods in which they accomplish similar tasks. We need to give them more of the same opportunities in captivity.

THE TWENTY-FIRST CENTURY ZOO

Parts of this book have explored some of the difficulties encountered in attempting to bring nature to captive facilities. It should be clear that the passage of two decades is unlikely to overcome all of the pressures against "pure" institutional nature study. Real predator-prey interactions will only slowly be introduced in the public education and recreation sections of zoos and wild animal parks. Few zoos will provide the solitary natural habitats that many species prefer because the technology required to provide apparent privacy for the animal while allowing visitor observation requires careful design and considerable effort. There is little evidence to suggest a change in the frightening rate with which *Homo sapiens* continue to eliminate species and greedily consume more and more of the unreplenishable parts of our world. Consequently, zoos will probably be repositories for endangered and threatened species as long as they continue to exist. Natural exhibition in which untreated disease, rejection from the feeding group, or struggles of dominance may eliminate some of these survivors is not likely to be tolerated by regulatory agencies responsible for their welfare.

All of these predictions suggest that for the indefinite future, behavioral enrichment of captive environments will probably lean heavily on artificial approaches. In fairness to the resident animals, these approaches should emphasize opportunities that are as voluntary as possible. The entire question of the nature or existence of voluntary behavior in many species is one that has filled a number of books, and no brief discussion can satisfactorily address all of the issues involved. Still, it does seem clear that for each of the freedoms that we human animals most revere there has been effort on our part to find ways to make our environments responsive. Surely, other species need the opportunity to live in environments sophisticated enough to provide similar feedback and entertainment. People are no longer amazed that artificial exercise devices may help to maintain our health and vigor. We also recognize the necessity to teach our youngsters skills necessary to enjoy new opportunities. Yet, there is something that worries some people about teaching other species new ways to interact with and exploit their environments. Perhaps all of this is best understood by recognizing what Heini Hediger (1950) illustrated long ago: it is a human *misconception* that animals are free in nature.

Paradoxically, one sure route to maintained enslavement is continued emphasis on this apparition of freedom. I remember one person telling me that he thought that it was a shame that people had to work and that the last thing that he would want to do was to provide opportunities for "animal employment." He went on to describe how wonderful life would be if all you did was lay under a tree, have nutritious food brought to you, and have your medical needs looked after. The thoughtful reader will recognize this prescription as an ideal one to produce shortened and uninteresting life spans in most captive species. Equally important, as I have emphasized from the outset, if individuals kept captive in zoos are to be good representatives of their species' capabilities and beauty, this indolent model would be disastrous. Our increasing capability to produce natural-appearing exhibits and dynamic graphics must be coupled with efforts to encourage and ensure species-appropriate activities.

My vision for the twenty-first century zoo is, at the minimum, an honest educational conservatory in which animals gain our increased respect as we provide them more substantial opportunities. In the best case, zoos may become what they often express a hope to be: places where people may learn to love and respect the uniqueness of each species.

REFERENCES

Andrews, P. and Groves, C.P., 1976. Gibbons and brachiation. In Rumbaugh, D.M. (Ed.), *Gibbon and Siamang.* Basel: S. Karger: 4:167–218.

Bandura, M., 1974. *The effects of an operant conditioning experiment on the social behavior of a captive group of diana monkeys.* Paper presented at 54th annual meeting of the Western Psychological Association, San Francisco.

Batten, R.P., 1976. *Living Trophies.* New York: Thomas Y. Crowell Company.

Beach, F.A., 1955. The descent of instinct. *Psychological Review,* 62:401–410.

Beck, B.B., 1967. A study of problem solving by gibbons. *Behavior,* 18(1–2): 95–109.

Carpenter, C.R., 1940. A field study in Siam of the behavior and social relations of the gibbon *(Hylobates lar). Comparative Psychological Monographs,* 16:1–212.

Chasan, D.Z., 1974. In this zoo visitors learn, though no more than animals. *Smithsonian,* 5(4):22–29.

Cheney, C.D., 1978. Predator-prey interactions. In Markowitz, H. and Stevens, V.J., (Eds.), *The Behavior of Captive Wild Animals.* Chicago: Nelson Hall: pp. 1–19.

Chivers, D., 1972. The siamang and gibbon in the Malay Peninsula. In Rumbaugh, D., (Ed.), *Gibbon and Siamang.* Basel: S. Karger: 1:103–135.

Davis, R.R. and Markowitz, H., 1978. Orangutan performance on a light-dark reversal discrimination in the zoo. *Primates,* 19(4):755–759.

Denenberg, V.H., 1962. The effects of early experience. In Hafez, E.S.E., (Ed.), *The Behavior of Domestic Animals.* Baltimore: Williams and Wilkins: pp. 109–138.

Diamond, M.D., Krech, D., and Rosenzweig, M.R., 1964. The effects of an enriched environment on the histology of the rat cerebral cortex. *Journal of Comparative Neurology,* 123:111–120.

Ellefson, J.D., 1967. *A natural history of gibbons in the Malay Peninsula,* Ann Arbor: University Microfilms.

Essock, S.M. and Rumbaugh, D.M., 1978. Development and measurement of cognitive capabilities in captive nonhuman primates. In Markowitz, H. and Stevens, V.J., (Eds.), *The Behavior of Captive Wild Animals.* Chicago: Nelson Hall: pp. 161–208.

Gossette, R.L., 1973. Comparative analysis of serial discrimination reversal (SDR) performances of the gibbon, *Hylobates lar*. In: Rumbaugh, D., (Ed.), *Gibbon and Siamang*. Basel: S. Karger: 2:208-220.

Harlow, H.F., Uehling, H., and Maslow, A.H., 1932. Comparative behavior of primates; I. Delayed reaction tests on primates from the lemur to the orangutan. *Journal of Comparative Psychology*, 13:303-343.

Hediger, H., 1950. *Wild Animals in Captivity*. New York: Dover Publications.

Hickling, C.F., 1963. The cultivation of Tilapia. *Scientific American*, 208(5): 143-152.

Jouventin, P., 1973. Observation sur la socio-ecologie du mandrill. *Terre et la Vie*, 29:493-532.

Larsen, T., 1971. Polar Bear: lonely nomad of the north. *National Geographic Magazine*, 139:574-590.

Lehrman, D.S., 1953. Problems raised by instinct theories. *Quarterly Review of Biology*, 28:337-365.

Markowitz, H., 1973. *Biological and behavioral research with captive exotic animals*. Paper presented at the 81st annual meeting of the American Psychological Association, Montreal.

Markowitz, H., 1974. Analysis and control of research in the zoo. In: *Research in Zoos and Aquariums*. Washington: National Academy of Sciences, pp. 77-90.

Markowitz, H., 1975. In defense of unnatural acts between consenting animals. *Proceedings of the 51st Annual AAZPA Conference*, Calgary, Canada.

Markowitz, H., 1976. New methods for increasing activity in zoo animals: Some results and proposals for the future. In *Centennial Symposium of Science and Research, Penrose Institute, Philadelphia Zoological Gardens*. Topeka: Hills Division Riviana Foods; pp. 151-162.

Markowitz, H., 1978. Engineering environments for behavioral opportunities in the zoo. *The Behavior Analyst*, 1(1):34-47.

Markowitz, H. and Becker, C.J., 1969. Superiority of "maze-dull" animals on visual tasks in an automated maze. *Psychonomic Science*, 17(5):171-172.

Markowitz, H., Schmidt, M.J., and Moody, A., 1978. Behavioral engineering and animal health in the zoo. *International Zoo Yearbook*, 18:190-194.

Markowitz, H., Schmidt, M.J., Nadal, L., and Squier, L., 1975. Do elephants ever forget? *Journal of Applied Behavior Analysis*, 8:333-335.

Markowitz, H. and Sorrells, J., 1969. Performance of "maze-bright" and "maze-dull" rats in an automated visual discrimination task. *Psychonomic Science*, 15(3):257-258.

Markowitz, H. and Stevens, V.J., (Eds.), 1978. *Behavior of Captive Wild Animals*. Chicago: Nelson Hall.

Markowitz, H. and Woodworth, G., 1978. Experimental analysis and control of group behavior. In Markowitz, H. and Stevens, V.J., (Eds.), *The Behavior of Captive Wild Animals*. Chicago: Nelson Hall: pp. 107-131.

Meyer-Holzapfel, M., 1968. Abnormal behavior in zoo animals. In Fox, M.W., (Ed.), *Abnormal Behavior in Animals*. Philadelphia: W.B. Saunders: pp. 476-503.

Meyers, B., 1971. Early experience and problem-solving behavior. In: Moltz, H., (Ed.), *The Ontogeny of Vertebrate Behavior.* New York and London: Academic Press: pp. 57–88.

Mitchell, G., 1973. Comparative Development of Social and Emotional Behavior. In Bermant, G., (Ed.), *Perspectives on Animal Behavior.* Glenview, Illinois: Scott Foresman and Company: pp. 102–128.

Morike, D., 1973. Verhalten einer gruppe von dianameer-katzen in Frankfurter Zoo. *Primates,* 14:263–300.

Mountfort, G., 1973. *Tigers.* New York: Crescent Books.

Myers, W.A., 1978. Applying behavioral knowledge to the display of captive animals. In Markowitz, H. and Stevens, V., (Eds.), *The Behavior of Captive Wild Animals.* Chicago: Nelson Hall: pp. 133–159.

Neuringer, A., 1969. Animals respond for food in the presence of free food. *Science,* 166:339–341.

Paquet, P.C., Markowitz, H., Sullivan, J.O., and Bragdon, S., 1979. *Observations of pack dynamics and mutual rearing of simultaneous litters in the wolf, (Canus lupus).* Paper presented at 59th annual meeting Western Psychological Association, San Diego.

Rowell, T.E., 1971. Organization of caged groups of *Cercopithecus* monkeys. *Animal Behavior,* 19:625–645.

Rowell, T.E., 1972. *Social Behavior of Monkeys.* Baltimore: Penguin Books.

Rumbaugh, D.M., 1965. The gibbon infant, Gabrielle: its growth and development. *San Diego Zoonooz.*

Schmidt, M.J., 1978. Elephants. In Fowler, M.E., (Ed.), *Restraint and Handling of Wild and Domestic Animals.* Ames: Iowa State University Press: pp. 709–752.

Schmidt, M.J. and Markowitz, H., 1977. Behavioral engineering as an aid in the maintenance of healthy zoo animals. *Journal of the American Veterinary Medical Association,* 171(9):966–969.

Schmukal, G., 1974. *The effects of feeding techniques upon the behavior of captive servals (Felis serval).* Unpublished BA thesis, Reed College.

Scruton, D. and Herbert, J., 1972. The reaction of groups of captive talapoin monkeys to the introduction of male and female strangers of the same species. *Animal Behaviour,* 20:463–473.

Soper, E.T., III., 1974. *The social behavior of captive diana monkeys under temporally dispersed and temporally localized feeding conditions.* Unpublished Master's thesis, Pacific University.

Stevens, V.J., 1978. Basic operant research in the zoo. In Markowitz, H. and Stevens, V.J., (Eds.), *The Behavior of Captive Wild Animals.* Chicago: Nelson Hall: pp. 209–246.

Stirling, I., 1974. Midsummer observations on the behavior of wild polar bears *(Ursus maritimus), Canadian Journal of Zoology,* 52:1191–1198.

Struhsaker, T., 1967. Behavior of vervet monkeys and other cercopithecines, *Science,* 156:1197–1203.

Tapp, J.T. and Markowitz, H., 1963. Infant handling: Effects on avoidance learning, brain weight and cholinesterase activity. *Science*, 140:486–487.

Wemmer, C., Von Ebers, J., and Scow, K., 1976. An analysis of the chuffing vocalization in the polar bear *(Ursus maritimus)*. *Journal of Zoology*, 180:425–439.

Wolfheim, J. and Rowell, T.E., 1972. Communication among captive talapoin monkeys *(Miopithecus talapoin)*. *Folia Primatologica*, 18:244–255.

Yanofsky, R. and Markowitz, H., 1978. Changes in general behavior of two mandrills *(Papio sphinx)* concomitant with behavioral testing in the zoo. *Psychological Record*, 28:369–373.

Yerkes, R.M., 1925. *Almost Human*. London: Jonathan Cope.

Tripp, J. and Mack, W., R. (1961) Inheritance Diet, Effects on voluntary activity, learning, mean weight on multigenerations. Psychonomic 14(4a) Dev.

Werner, G., L. Harris J., and Seas, H., 1970. Analysis of the influence of socialization in the mouse on aggression. American J. of Zoology, 130, 416, 490.

Whitman, H. and Rowell, Tm., 1974. Control after among apes chimpanzees and Orchard reacting Zool. Primatum no. 19, 254-255.

Vandier, R. and Stackowski, H., 1974. Changes in sexual behavior in two mammals following castration. J. comparative Phsyc. Psychiatric castrate in the sex Psychoendocrinol reprod. 38, 253-257.

Yerkes, R.M. 1925. Almost Human, London, Jonathan Cape & Co.

AUTHOR INDEX

SUBJECT INDEX